THE PLANETS

THE
PLANETS

A
COSMIC
PASTORAL

by
Diane Ackerman

William Morrow and Company, Inc.
New York   1976

Copyright © 1975, 1976 by Diane Ackerman

The "Full Moon" section of the poem "Earth" originally appeared in *Poems* by Diane Ackerman, Jody Bolz & Nancy Steele, published by Stone-Marrow Press, and is copyright 1973 by Diane Ackerman, Jody Bolz and Nancy Steele.

Grateful acknowledgment is made for permission to reproduce the following:

Frontispiece: "Nackte Weibliche Gestalt, Den Tierkreis Haltend" by Albrecht Dürer, reprinted from *The Complete Woodcuts of Albrecht Dürer*, ed. by Willi Kurth, Foyle, London, 1927.
Photograph of Mercury   NASA
Photograph of Venus   NASA
Photograph of Earth   NASA
Photograph of Full Moon   NASA
Photograph of Cape Canaveral   NASA
Photograph of Mars   NASA
Photographs of Phobos and Deimos   NASA
Sample of Martian Nomenclature reprinted from T.L. MacDonald, "The Origins of Martian Nomenclature," *Icarus* 15, 1971.
Asteroids reprinted from William K. Hartmann, "The Smaller Bodies of the Solar System," *Scientific American*, September, 1975.
Photograph of Jupiter   NASA
Photograph of Saturn   NASA
The Moons of Saturn   Arlene Goldberg
Photograph of Uranus   Lick Observatory photograph
Christie's Diagram reprinted from G.L. Siscoe, "Particle and Field Environment of Uranus," *Icarus* 24, 1975.
Photograph of the comet Kohoutek   NASA
Photograph of Neptune   Lick Observatory photograph
Photograph of Pluto   Hale Observatories

All rights reserved. No part of this book may be reproduced or utilized in any form or by any means, electronic or mechanical, including photocopying, recording or by any information storage and retrieval system, without permission in writing from the Publisher. Inquiries should be addressed to William Morrow and Company, Inc., 105 Madison Ave., New York, N.Y. 10016.

Printed in the United States of America.

1   2   3   4   5   6   7   8   9   10

Library of Congress Cataloging in Publication Data.

Ackerman, Diane.
    The planets: a cosmic pastoral.

    I. Title.
PS3551.C48P6   811'.5'4   76-14840
ISBN 0-688-03088-2

BOOK DESIGN   ARLENE GOLDBERG

DIFFRACTION

for Carl Sagan

When Carl tells me it's *Rayleigh scattering*
that makes blue light, canting off molecular

grit, go slowgait through the airy jell, subdued,
and outlying mountains look swarthy, or wheat

blaze tawny-rose in the 8:00 sun, how I envy
his light touch on Earth's magnetic bridle.

Knee-deep in the cosmic overwhelm, I'm stricken
by the ricochet wonder of it all: the plain

everythingness of everything, in cahoots
with the everythingness of everything else.

The second pair of pants in my genetic suit
held no whys and wherefores, no clement unity,

no federation of water-flea and Magellanic cloud.
Mathematics is a language I don't speak.

I can't unveil the sun's ricy complexion, really
fathom Vela-X ticking like a clock, track comets

on the run through hyperbola, parabola or ellipse.
I'm bone-deaf to cloud-chamber music.

When Carl tells me it's *Rayleigh scattering*
that azures the sky or unpuzzles rainbow-

grinding weather, I envy his firm grip
on a world where I think not as a thinker

thinks, but as light engrossed in every object:
a doting consciousness among alien forms.

I only know, one rural twilight, when wheat
blazed like ambergris and a chicory sun

haggled with a black sky, for a moment
all the blues of the world scattered;

my ribcage sprang open like calipers, and,
in their widening compass, nothing lacked.

ACKNOWLEDGMENTS

I am grateful to the editors of the following periodicals for first publishing some of these poems: *Carolina Quarterly, Garfield Lake Review, Granite, Harvard Magazine, Massachusetts Review, Poetry Now,* and *Sito.* "The Other Night" was awarded the Abbie Copps Poetry Prize (Olivet College) in 1974. "Uranus" received a Heermans-McCalmon playwriting prize in 1976. Some parts of *The Planets* appeared in Carl Sagan's *Other Worlds* (Bantam, 1975). Selected readings were heard on the CBC radio series "Ideas," broadcast in 1975. Most of these poems were written while I was being supported by a Cornell-Rockefeller Fellowship in Humanities, Science and Technology; I am grateful for the time thus provided me. Over the past year or so, I have put all too many questions to the Space Sciences faculty of Cornell University (especially Carl Sagan, Shirley Arden, Joseph Veverka, Alan P. Lightman, and David C. Pieri), and I thank them for their generosity of time and spirit; any errors in this book are in no way attributable to them.

CONTENTS

PROLOGUE

While a blue-green tea-clipper
called Earth lists starboard
in the bowels of summer,
on the birthday of Copernicus,
I'm beginning this poem
with what I plan to ignore—
the shuntling of the white tulle curtains,
beyond which day-lilies
are breaking out like sunspots;
the platoon of cream and caramel bugs
turning the yard into a leggy twitch;
the rodents with their tiny eyes
black as sen-sen; the Lackawanna
wrapping its wet thighs around a delta;
even the woodpecker typing
on my sun-broiled shutter—
all the bud-breaking, polychrome
jibber-and-fidget, volcanic mayhem
barreling through summer. Whose weedy
delirium I quit, but reluctantly,
and with other deliriums in mind:
to pull free of Earth's whalebone stay
and sally out onto the cold compress

11

of the universe, lending my psyche
to Copernicus for a commemorative issue.

Perhaps I've too much at stake
on these cryptic Petri dishes orbiting
the Sun. If nothing's afoot here
but dumbbell, loveseat, and oarlock craters,
hotbed simoom and marmoreal dust,
in what clinic will my dashed
hopes renew? Sweet Urania,
line your airy lungs with gold,
and bear with me a while.
I'm young as I write this and green,
yet in my lifetime we'll never
sail beyond Pluto, or cut time
on the bias in a black hole in space,
even leave the twirl of woodash
that's our Milky Way. For me,
the Crab Nebula will never
be made real, so I'm lighting out
for the planetary wilderness,
a gambler for whom it's either
surfeit or famine. The Planets
are nine dice rolling in the dark.

MERCURY

MERCURY

A prowling holocaust keeling low in the sky
                                    heads westward
for another milkrun.
The Sun never sets on the Mercurian empire:
                                    it only
idles
on each horizon and lurches back, broiling
                                    the same arc
across the sky.
Day in and out. A target gone berserk
                                    in a shooting
gallery.

                        On the bright side,
Mercury is hotter than a smelt-forge in the
                                    Kalahari
and smoldering
like the Devil's ass. Fitful lava-yolk spews
                                    up white

geysers,
each one braising, tentative, and self-
                                    willed,
leaving the mantle
a sheet of volcanic hives as icy-hot, some
                                        claim,
as the darkside's
permafrost is molten chill: both the fiercest
                                    luau-pit
in the solar system
and the coldest oubliette—all in one freak
                            Zoroastrian ball:
a plummeting
mine-field that whorls round yearly.

Anyway, that's the theory.
A wrong one. As it happens, every 30 days,
                            the planet turns
another cheek
to the Sun, ornamenting the night with its
                        moonlike phases.
Venus is much hotter.
Neptune colder.   Like our own terrestrial
                                        desert,
the topsoil's
knurly with mosaic grains that roundelay
                                the heat flow,
loath
to conduct. So the days are warm and the
                                evenings cool,
relatively
speaking: 940° Fahrenheit in the Sun and
                            -180 in the shade.
Hardly brochure weather.

                    But it wasn't always
                    so. Millennia back,
the Sun
broke the spirit of Mercury's axis through
                    Herculean
spinal drag.
A frothing ball once, double-timing on the
                    hop and unstoppable,
Mercury
wound down to a breezy mambo at first, then
                    adopted a tune
with new reticence.

                    Hob-nobbing
around the Mercurian noon, with only a thin
whiskey-colored veil
to waylay it,
the Sun must burn carte-blanche in the sky,
                    growing white-hot
                    and even larger
as the day grows, spitting, burgeoning, and
flaring out
at the seams
like an infra-red sunfish gravid with roe.

                    As preamble,
the Sun incommodes
all right. Cutting too broad a figure in
                    a diabolical sky,
its lesion face
peels off, melts down, and recreates itself,
                    metamorphosing

with vaudeville stamina:
Slowest poker-hot tennis ball on record.
                                    Shuttlecock
warping
the celestial loom. Oozing and malodorous
                                          wound.
Radioactive
brimstone- and lava-lamp. Windbag, rapid-
                                firing buckshot
torpedoes.
Lethal fenful of clarified pus.   Blazing
                                    white parrot
at red alert
Ulcerated dollop of creamery butter. Light-
                                       stutterer
snagged
on the letter D. Welter-weight. Suppurating
                                ointment ball . . .
and on and on
until the last avatar—a single cormorant
                                      with eye-
branding
plumage—calcifies into heaven's decoy.

                                    A man would
knuckle under, half-hatched by the rays
percolating down
through each blood
stratum, by the universal cannibalism and
                                       box-lunch
                                      of the Sun
shimmying pell-mell down the optic nerve,
while the brain turns
to ariose confetti,
packing it in, cell by cell. An autopsy
                                    would hedge:
                                    asphyxiation,

sunstroke, heat prostration: all mumbo-jumbo
for an overdose
of light, too many
pinwheels, hecatombs, and fandangos, when
                         bland form
                    should upwell,
like an exo-skeleton, be stable & unassuming
as a coat of primer,
a kind of
pitstop, backdrop, armrest, alibi, not this
                         crazy-quilt
                    that leaves you
hanging by your throat.

Plumping up,
the Sun will forcefeed color down brains
                    that doubletime
even in the dark. For what end, I can only
guess:
to pâté the cranial liver. Were they rooting
                         for truffles
in the Garden of Eden?

Twilight. An ochre slug,

                    mummified
and gargantuan, leaves no room

                    for lodestar,
horizon,
nightlight, or celestial blackboard with
                    its ciphers
erased,
no sensual winter, when Nature's dull ruff
                    cants the mind
indoors
to ruminate a while untainted, neither chance
                              nor

inclination
to reflect. Lambast, bludgeon, bombard, and
titillate.
You'd need
to cultivate boredom like a crop of Brussels
sprouts,
learn to
throttle the solar-fit just to think, and
end up
raving, avid
for a treeknot, sinkhole, grotto, sunspot,
or even tiny
bulbs of bacteria
on clover roots that store nitrogen cool in
the mocha dirt.

Shark
in the cosmic ocean, Earth's forever cruising
even when it
eats or sleeps:
a marbly denizen, a waltzing Diogenes.

So, if nothing
lingers on Mercury but heat (and even that
outgassing
and boiling off),
how will it matter? Us gandy-dancers know
what it means
to be in transit,
hog-tied, and yet somehow always on the run.

Mercury we'll untidy
for an open-air forge, bakery, glassworks
(but *people*

will be in ovens
cooling off, while kilns shunt freely top-
side).
We'll confect
puff-pastries & foamy bagatelles, handblow
crystal
ribboned
with melted lead, build dynamos to nurse on
pure energy
teats.
We'll be herdsmen: sunlight undiluted by
air's
the ideal
breeding ground for pit-vipery reptiles. Many
snakes
detect prey
through built-in infra-red receptors, not, as
you'd guess,
with their
divining tongues. And, as penal colony,
Mercury is
the next best
to Hell: a planetary Australia; police dogs
will give way
to anacondas
and pythons. In each patrol-tower, a hyped-up
snake
will be on the nod.

VENUS

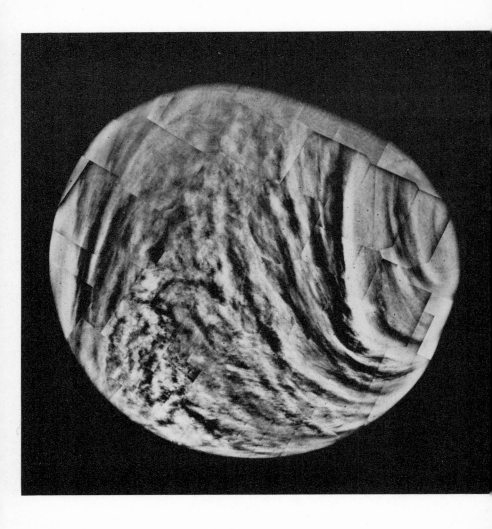

VENUS

Low-keyed and perpetual,
a whirling sylph
    whose white robe stripes
            around her; taffeta
wimpled like a nun's headcloth;
        a buxom floozy with a pink boa;
            mummy, whose black
sediment desiccates within; wasp-star
    to Mayan Galileos;
                an outpatient
wrapped in post-operative gauze;
            Cleopatra in high August—
    her flesh curling
            in a heat mirage
lightyears
        from Alexandria;
    tacky white pulp
spigoted
        through the belly of a larva;

                    the perfect courtesan:
obliging, thick-skinned,
                    and pleated with riddles,

Venus quietly mutates
          in her ivory tower.

                    Deep within that
                         libidinous albedo
temperatures are hot enough
          to boil lead,
     pressures
          90 times more unyielding
                    than Earth's.
And though layered cloud-decks
                    and haze strata
          seem to breathe
                    like a giant bellows,
heaving and sighing
          every 4 days,
the Venerean cocoon
          is no cheery chrysalis
brewing a damselfly
          or coaxing life
                    into a reticent grub,
   but a sniffling atmosphere
40 miles thick
          of sulphuric, hydrochloric,
          and hydrofluoric acids
all sweating
                    like a global terrarium,
          cutthroat, tart, and self-absorbed.
No sphagnum moss
          or polypody fern here,
          where blistering vapors

and rosy bile
        hint at the arson
with which the Universe began.

                Topstory, the dewy
cloudbreast purls,
        oddly Earth-like: room-temperature,
                        1 atmosphere,
lots of sun . . .
        a no man's land
where, one day,
        archipelago space-labs
            (called "aerial sleuths")
                    will string along
        like Japanese lanterns
            gaily bobbing
in the Cytherean pink,
        while telepuppets revive
        from each corrosive dunk;
meteorologists
        fly cybernetic windsocks;
                        terraformers,
    venting micro-organisms,
shake down
        the planet's inhospitable air;
and a rapture of exobiologists
        perches, hymning
                for life-sign water and oil:
Venus in one of her aqueous humors.

But how drab the vista:
            ruddy pink fog
        tonelessly milling, and rumored planet,

invisible
  to the naked eye,
              somewhere below
idly rolling a haunch.
              At wit's end,
stupefied
  by the grueling blandness,
              prey to claustrophobia
and puncture psychosis,
          even veteran starfolk
              will potter about
drugsome,
      unless artisans
    weave great litmus banners,
              forge metal murals
whose designs
    unfold as they corrode,
          or interface the air
  with holographic pictures:
          one week
    an hyena-littered savanna,
and the next
  a tireless Bolshoi ballet.
          "Have you seen the *floats*?"
    they'll inquire,
meaning
  no rose-strewn trucks
              paraded dully
      through holiday streets,
  but a full-blown walking-tour
      of the senses:
cloud-anchored armadas of merciful art.

          Could one override
the instinctive urge
          to pity where nature is pitiless

(and land *coup de grâce*
on a planet woefully hemorrhaging below),
             could one elude
the bends
      (as if down 3,000 feet
                  ocean-deep
                  while a gas envelope
chokes tighter and tighter),
             could one stand
the heat
      twice a household oven's
                  midnight and noon
             both
        steeped in vermilion
                  or glassy scarlet
      (even shadows
             red-hot from their own fire
upglowing like sudden bruises),
             could one weather
the equatorial trade-winds
                  scarcely Tahitian
        which gust 200 mph
and pan out to cyclone-cap the pole
        (their ballistic undertow
   sucking and howling),
             he'd find
        no global oilfield
                  smoldering like tallow,
no blustery desert,
   no seltzer ocean,
no playa of detonating volcanoes,
                  no primeval swamp
        braided with coal—
in fact,
   none of those journal wonders
        Venerean lore
seems to be storied with—
                  Venus neither in furs

nor attached to her prey,
but gamy, flat-chested,
       and covered with scurf.

      Outlandish
          though it sound,
Venus is a kind of planetary midwest
    pockmarked
      with long shallow craters
  where meteors bit
      into her pliant flesh
leaving
  their crude vaccinations
    as calling cards.

But who'd notice?
  Dilute in the lugubrious gloom,
    light's broadcast
  so many times
that color happens
for only tens of feet
    and panorama not at all.

The sun rises in the west
    as though disinterred:
  a faint garnet haze
      whose coming and going
  torches the horizon
like a dingy red sash,
    churning out
      optical hoodwinks
by the score:
    one's own hand,
      outstretched,
  would float above the arm
  like a myopic god

stymied
    by the wide pink yonder,
  one's sprawly feet
      would hover
     like sycophantic trout,
or gingerly parrot and dip
   waging
an egotistical mating dance.
    Suddenly batched
      by the charnel twilight,
  a single man
   could be arrested
for unlawful assembly,
      while everything
    on and just past the horizon—
mountains; craters;
space-garbage;
      billboards;
   the supple planet-skin
  healing itself;
alfresco suicides
   becoming one with nature
  as each fragile life-molecule
      shakes itself apart;
choirs of self-anointed shamans;
   political candidates
  jockeying for stature;

       gore
  cast by a holographic prankster;
   funeral groups
 scattering bodies (not ashes)
which immediately precipitate
    out of themselves in ultimate,
if somewhat literal,
   ecstasy;
    Dante seminars—
the whole putrid, ghoulish stew,

boiling overhead
like a mortal haggis,
would form
a conical epiphany
that funneled
straight into the brain.
Grey matters, indeed.
Tourists planning
a Venerean junket
should bear in mind
that for schizophrenia
you needn't keep a day free.

Unsullied by moons,
radiant and blowzy,
aging
with almost no midriff bulge,
Venus trills
her apothecary cloud:
a seasonless queen
rolling each languorous hip
so slowly
no magnetic web forms at all,

or a turbulent spangle,
shimmery and dead.

Either way,
the planet tilts
the same face
Earthward
whenever she's near,
her pearly tonnage
swinging
like a giant compass
newly risen to our iron heart.

EARTH

EARTH

I REVEILLE

This morning
ice hung on the shutter
like a milky tusk.
From a neighbor's chimney
smoke rose grey as the sky.
And I was full of stagefright
and misgiving. How shall I
celebrate the planet
that, even now, carries me
in its fruited womb?
Could a fish, locked
into its salty element,
know about Earth
where seas high-tide
and the oceans are part
of an atmosphere
they evaporate and rain

through, or that the oceans
are part of the landfall
when they freeze to ice-cap
and glacier? Could a fish,
blinded by water, know
of the tinkling gold
lights on the trawlers
anchored off Port Canaveral,
how the tide swarms in
like a herd of rodents?
How can any system
observe itself? Should I
try through catalogue
or epitome? Let me begin
with a small cosmogony,
and then cut a swath
through this rabid planetary
life that, for a strike-root
moment, I was born to.

II COSMOGONY

Long ago, Earth bunched its granite
to form the continents, ground molar
Alps and Himalayas, rammed Africa
and Italy into Europe, gnashing
its teeth, till mountain ranges buckled
and churned, and oceans (salty once
rivers bled flavor from the seasoned
earth) gouged their kelpy graves.
And the rest is history: Earth
tugging up its socks each night

(even the solid land ebbs and flows),
oxygen and nitrogen spat by biology,
limestone built from tiny sea-creatures,
floppy fish revising from pond to pond,
tree-shrews, apes, painted cloud mesas,
the red thresher in the cornfield
stiff-necked as a cockerel.

III EARTHSHINE

Mars and Jupiter stud the sky with light.
I watch them nightly, and try to understand

that I am on a planet, a *planet*, like they are
planets. I think of Mercury, pockmarked

by the Sun's yellow fever, of that flossy
white node in the galactic marrow called Venus,

of Saturn with its pussyfoot ice,
of cyclops Jupiter in a pinstripe suit,

whose pearly moons float like bons mots,
of Neptune, whose breath is ammonia,

of gangrene Uranus, ghoul of the heavens,
of Pluto, rock-ribbed as a die-hard comet.

But what vision could bridle my own
Earth-planet, so headstrong and diverse?

I look out to see
what the broadleafed evergreen
and chickadee
        are making of the weather.
If the birds puff fat,
        it'll be in the 30's.
If rhododendron leaves
fold like praying hands,
        much icier.
Like this planet,
I'm full
        of useless information;
for example:
Galileo,
        contemplating the Earth,
once muttered under his breath,
        "It moves."

Wrapped in a light-blue shell,
Earth croons air and ocean color
like the egg of some extinct bird
left to ripen in solar heat,
its jelly thick and mellow.

Blinding white clouds rally
and sprawl through tufted fleece
and high patchy swirls that blur
the whole planet
rolling beneath them like a code.

But here and there, through hazy
cloudgaps, the oceans and continents
blink their pastels, tingeing
gaily into one another
all their hard divides.

From afar, no human ken
or browsweat comes to light,
only a deluxe planet,
crop-happy as a citadel, bustling
behind its frigid black moat.

IV INVENTORY

Beaver stipple,
ice mantle, belly skid;
tea-kettle steam
coiling like a hindu rope;
Australian woodlice
ticking marine iguanas;
rattle-knobbed wild onions;
pale petaled pickotees;
cicadas sun bathing
to heat-prime their muscles;
rex begonias abristle
with cat-raspy tongues;
frogs croaking in dialect;
white bottle gentians;
boll weevils insinuating
their lives through dirt clods;
the wasp's internal
hourglass, light-flipped
to clock summer;
amphibian bones gemmed in strata;
the corm of a crocus;
my grandmother, with long braid
and alabaster skin,
shedding her life
under a cheek of stone.

I have so much
invested in your Earth,
whose dust I was born out of
and will bleach into.
And yet I'm lame to sing
of all the cloud tufts,
the rivers and oceans
and aprons of land,
the volcanic spasms
and the crimped sierras,
the plants and animals
and, above all, the motion.

Imagine, we live in a world
so riotously packed
with buzz, bloom, burn and fidget,
we actually tend
to find *Quiet* freakish,
Calm ominous, prolonged
Stillness death-defying.
Why, you'd think
one would never cotton
to anything, never grow
bored, not succumb to habit,
but only craze slowly
from terminal surprise.

A queer lot, aren't we,
on this rickety oasis:
whirling men
on a whirling planet,
whose organs slosh
right along with the seas:
4 billion salt-licks
of muscle and blood
dissolving in one prominence
of one sun. As though
life were motion unrelieved.

V RAPTURE OF THE DEEP

As I walked toward the scuba port,
watching the ocean—all its tinsel
crackling in the noonsun
as if half the galaxy descended—
I imagined bass loping wall-eyed
through each coral thicket, dolphin
plying the air's open hollows,
a twill of sturgeon cruising
below the flat soapy tide.
The Caribbean blued like enamel:
one unbroken surface and periphery.
And I thought, never again
will I lip-read you like this,
find your spineless hinterland
gaping and unplumbable.
Well-baited as a stickleback,
I'll enter your thick skin
and watery limbs, glad-eye
your rippled bottom, kick up
a little dust. And no longer
will you fret me, icy and withdrawn,
your green body silent as a jell.

When the ochre beach withered to a thread,
I leaned over the depth-haunting water,
my hands white as salt-cod. The boat cradled
on its makepeace keel. And I thought,
how like a diamondback the sea lies dozy,
everywhere glutted from a recent feed.
I tugged my flippers on, then rimed saliva
round the maskplate to defog it,
swilling saltwater overall. And I thought,
how light blisters the fluted sand below

when, not ten yards off, the floor cants
tea-brown to gunmetal grey, then blackens,
sabled as a friar. The guide catechized me
in scuba-talk: *Are you OK? I'm OK. Up. Down.*
I'm in trouble. Nothing major. Go that way.
My ears haven't cleared yet. Danger, over there.
And I thought, what rum patois can this be,
in whose clipped handspeak we all signal
our muteness, as unable to sign color as joy?
If my heart flattens like a sand-dollar,
its flocked breath too quickly vent, what sign
will tell you fear coils in my joints, melting
the cartilage away? And if my heart blooms
like a sea-fan, daft with each tidal undersong
and coda, what sign will tell you how my bloodwings
beat, what sign how my vein-harp rings?

Dave rocked the aqualung to my hip,
then tightened the girth, till
its ice-flow, chill-glazing my spine,
frosted each pillar away.
For this, my blood would strut back
heartward, shabby, addled, well-spent.
So, leaning in to its airy bother,
I hoisted my life upon my back:
the 70 minutes of its ride-hard gallop
and the 71 of its end, fluent
in a single vat; unwieldy, it pulled
through the head and shoulders.
Then, saddled with my life,
I careened gunnel-long, bracing myself
on the boat-smooth acrylic,
only just thought to lean when gravity
prospered and, suddenly, as if
through a trap-door, I tipped open
like a flower, and fell in with the sea.
I remember the moon, out full

in the daylight, flashing by like any other
round cloud, the soft elision
of water mounting my airborne thigh,
the unhurried tumble, the swift upreel,
how buoyant my life had become.

Floating downward, as the pressure bulked,
a white-hot cable swizzled through my brain.
Ears, I tapped, *My ears*. The guide lulled me
upward barely a foot to where pain dropped
like a whim. Again, as we lowered, the inner
ear blazed: gravity's watermark on my vellum,
or ill-begotten evolution (what gene now
harbors so landless a gift?), then the lilt
up and the pendent wait. Fluids, crackling
in my ear, let fly. *Are you OK?* My hand shadow-
puppets *I'm OK*, as we downfall in thickening
quiet, amplified by each lung-wheeze. Inhalation,
exaltation. Sounds of life and life escaping,
then the coral reef thrown open like a dream.

A Grunt sashayed through the water, its minnow-herd
pearly in the abstruse sunlight. Creviced or out stark
on the coral butte, sea-urchins' black thistles
ticked like burrs under the ocean's saddle, venom
locked in every chamber. Where a snail left its ivory
hull, a tiny tusk lay coated in the largo seabed.
Long white anemone fingers drummed, replete with cilia.
Wide-eyed Grey and Yellowtailed Snappers ambled by.
A Harlequin-bass paraded like a freckled buckskin.
Pollen-bellied Tobaccofish zig-zagged through the mesh
of a junked box-spring. Royal Grammarians looked twee:
all purple robes and golden butts. Damselfish
nicked their velvety black heads. Tinted
Pearjacks hobnobbed with Sand-tiles. Like a wedge

of tuttifrutti sherbet, a Parrotfish grazed
organisms off the reef. A Porcupinefish drew up,
readying its goiter. I peered around the algal haze,
everywhere endless and full. The sea,
giving up the ghost of color, now prismed light and dark.
Had Caravaggio been here with his under-ink, painting
chiaroscuro from the whaleroad up? Drug-thick
in the holdback water, I felt rubbery, and slowgaited
till each pore bristled. Touch kicked on
like an old furnace. A museum of flowery paperweights
domed below, where live, thin layers of coral eluded
the bloodsucking starfish horde. Brain-coral
ribboned like disembodied minds. A clam opened
its sandy locket: the apparent flex witless as a tic.
And yet, at once, my heart reeled in its bony cage,
dithered and shook till I thought its blue seams
would rip, then spun agog, wild to blurt its wonder.
And I said, yes, *yes,* you are a bivalve, too.

In the boat, my lungs taut as pufferfish,
and hands dry-iced to the gunnel, I'd watched
the guide, an ex-druggist smitten by the sea,
tread gaily above the parchment sand
where he landed in that ocean on his feet.
David Heath, I dreamt your dream last night:
flying off the lip of a marine canyon,
sailing open-limbed over trenches, arroyos,
gingerknobs, and plummets: a gypsy manta
light-crazed as the algal hillocks, freefalling
till nitrogen narcosis took my breath away,
luring me back into its coral braintrust,
deep into voluptuous ether (like my womb
smelling of herringbrine), but gill-less now,
lame to feed and filter. Mechanically, I stroked
my gills, gone in the pallid inland flows,

gone in the plungings of water and wind
pummeling the shore for a single grain
to carry in a pocket of sea. My blood eddied
with the night-tide on patrol; my lungcork swam.
And, tossing off my reason like a second sweater,
I paddled the very water the sea-squirt did,
cousin of the lobe-fish (humble make-do
that in a pinch coined backbone, lungs, and feet),
cartwheeling past diatoms and flat-water crabs
bronzed for posterity by clay and sand, pell-mell
through leagues of undefiled chaos, every sac,
airhole, and pore cued to gravity's hug,
till little stood between me and absorption: only
one papery 1/16 inch: those *ad hoc* billennia of flesh.

Low smothery cloudbanks have been weeklong
descending, as if the snow-broken world
were Desdemona. And I wish they'd get on
with it: drop their load, snuff the holdout
beetles and frogs, wind-rip the dry seedpods
tree-clinging with all the tenacity
of saints, oblige the wall-ivy to shed
its ripe skin, muscle under the ice-hipped
weed. Whatever winter does, let it do it
already, and cut the shilly-shally.
Even the silver birches look addled.

I stood over the floorvent, and let
the building's hot breath pour up under
my nightgown. Outside the window,
a Chevy, fender-deep in snow, flung up divots
like a dog covering its fecal tracks.
I drank my tea and began the dishes, dipping
each fork and amber glass idly in warm suds.
I scoured the pot, ran a sponge around
the saucer rim . . . and my heart thickened.
At seabottom, I'd gripped a honeycomb,
puzzling its cratered hide till it burgeoned,

then squooshed ever so gently,
and, as its gray effluvia colored my hand,
felt my bloodtree ripen like a conifer.
The indisputable sweet alarm ran through me
like a spark. *It was a sponge, a sponge.*

The sinkwater gurgles down ceramic
sand, weaving as the Pearjack had, dumb
to the cackling horns above, arboretum
of streetlamps, peach-colored Arcturus,
fully blind to the snow-ready sky.
I found a niche there, too.

VI VESPERS

Oh blessed *E. coli,*
florid in my loom; stomach acid, hotter
than a dung-beetle's
belly; mogul enzymes, butting in on each
chemical transaction;
ropy jams and jellies; filthy leucorrhea;
scrimshaw teeth
the dental hygienist cleans like a cattle
egret; volleyball
cells dividing; muscles that tear like
english muffins;
caraway adrenals (bagmen for the steroids)
who rocket me
punchdrunk and ablaze, run me ragged,
cover me with down;

Oh silky white ovum
sweating out the new moon; cheesy vaginitis;
        russet nipple, taut
as shelf-ice; hormonal gendarmes, in my pyschic
        landscape, like angels
pouring out the 7th vial; every lugnut aspiration
        and nocturnal remission;
all the ugsome, virid, and bilious pomades;
        brainstem, thick
as rhubarb; moon-white fleshy scar on my thigh,
        nomadic cancer,
groping along the uterus with fell intent;

        Oh heart, in whose
china-red glaze I feel the *tsk tsk* of every
        pulsar; mangy aerobic
and anaerobic bacteria who, foraging, spread
        meningitis, typhoid,
abscess, gonorrhea; bone-wings of my pelvis,
        flaring to spare
the torso pain; hairlike flagella with your
        tiny spurs; salt-flat brain,
fatigued from the process called identity,
        in whose Petri dish
even mold is a godsend; nerve-net, shaking
        hands with a jolt;
the flick-flack synapses opening and closing
        like canal locks,
and then again the tiny leaps of faith;

        Oh neutral sodium,
I do not blame you; white corpuscles pawnbroking
        my blood; libertine
polyps, appliquéd onto the merest cell wall;
        emotional hemophilia;
radical neurons, always sated or recovering;

tiny mitochondria
powerplants; catalytic proteins, engaging like
a clutch; skeletal
polymers; islands of Langerhans rumming my blood
with your sweet treacle;
water molecules, shaped like glisteny croissants;
freckles, corns,
beauty marks, scars, and all the other amendments
to my constitution;
DNA backbone of sugar and phosphate—sweetness
and light—
magnanimously handing round your blueprint,
though you stinted me
the eyes I wanted: pale amber ones like two
peach flans;

Oh ringworm, thrush,
athlete's foot, and all the other fauna
of my disregard;
loaf-shaped, polyhedral, or spermlike viruses,
staked out even
in the smallest bacterium, who pass unnoticed
for 100 lifetimes,
then rage into a viral factory, devastating
and devouring its host;
kamikaze positrons; charmed quarks; as well as
the cutaneous and
subcutaneous epidermi, whose spongy jacket
my life wears . . .
I hymn you one and all. For, though I time
to time badmouth it,
I am truly grateful for this mind and body,
down to the last
wanton enzyme, particle, death-dealing cell.

VII PLAIN COUNTRY FOLK

Last night, I went to the horse auction
in Unadilla, a small rural town
upstate. There, just off the main road,
a rickety barn glared out of the blackness,
thick with people and livestock, trucks
and horse-trailers, mudtracks and manure,
hawkers and oglers. Inside, the air
was ropy with sawdust and smoke. Men chewed
twigs of straw and plugs of tobacco
to keep enough saliva in their mouths.
I can still taste the thick, dusty layer
of country life, full of sweat, horse
dander, rising hopes, and noise. Haggard
young women with dyed hair corralled
their children, and sat in the bleachers
like past lives, silent, watchful, humorless,
as their men dealt in the arena below.

Horsetraders crammed into the small enclosure,
each with a new story, or a tip, or a young
woman. One tells a string of dirty jokes
that aren't funny. Another how his son
shot him in the leg, how his coondog bit
a notch in his arm. Con-men gabble cautiously
among themselves, effusive, festive, giving
here and there an inch, here and there a wink.
Like gunfighters in the old west, they know
better than to take each other on, but every
now and then can't resist the sport of it.
Greenhorns, in clean pants and pastel hats,
weave through the crowd, ripe for plunder.
The lightning brogue of the auctioneer

49

warbles through yet another octave, sweeps
away a boxload of snaffles and hoofpicks,
then a dozen chaps and some Christmas toys.
A string of weedy horses goes to the packers
for 15¢ a pound.

We stroll through the warrens of ponies
out back. Each nag, with white label
and number on its tail, can be counted on
to have a little something wrong. This one
is a heaver, that one lame. A trader rolls
back the thick pink lips of a dun mare.
Children play tag in an empty box stall.
A few con-men deal squarely among themselves.
Shit no, my roan mare's a fucking weed,
I tell you; I got her pumped so full of bute
her goddamn teeth ache! But that app there
with the blaze is no screw. Come on,
give me three figures for him. . . .

Most everyone here is broke, or has been,
or will be: came with $5,000 last time,
and now has 28. A kindly old man,
with a grey wool cap and spreading hips,
looks woefully out of place, an easy mark.
He chats lightly to me with aged good will.
He's a pensioner come here for the circus
of it, some free entertainment in a lonely
and wretched life. I feel such granddaughterly
concern, and wish I could spirit him away
somehow, perhaps to a nice farm or country-
house, give the last of his dwindling life
a little substance. Out front, a breedy-looking
bay, who'll go lame tomorrow, is trotted
back and forth across the cramped arena,
her bad joints so full of aspirin
she sells high. The con-men almost smile,

look nimbly to one another, as tiny lines
around their eyes just crease.

Later, in the bar, drinking with traders
who are boisterous and bawdy and full
of good humor (no matter how their sales
went), I mention the old man in the grey cap,
and hear his cronies flood with laughter.
They retail his exploits for half an hour.
He is the biggest con-man of them all,
the toughest, the wildest, the most
resourceful, the one who can sea-change
through any age and outfit, with daredevil
speed and elaborate cunning. I'd been had
by the master, the king of the thieves.
Dumbfounded, I simply cannot believe it
of so pathetic and harmless an old man.
They say, *Sweetie, you waltzed straight
into the lion's den, and were luckier
than Job to come out in one piece.*

VIII   FULL MOON

Vampiric black ether
gnawed it last month: sickle, kayak, udder, wormy peach.
Then it grew
a bull's face and floret of cauliflower, turned polar-white
on its own
stark dawn. Tonight it's whole as wheat and cartoon-round,
a ball of sump

that swills water and makes babies drop. Look, I can push in
a floury thumbprint
and, everywhere, women will drain white as shallots, hemorrhage
in their vents
and fizzle off like gas. (I think of the stars as a million
grey cataracts.
It leaves me hanging by my eye-teeth, but it leaves me hanging.)

And now that
dazzling keyhole they call a moon turns out frigid as a bat,
a slab
of scar tissue good for nothing but plant food—better than
horsemeat,
but not so cheap, say, as dung. That marbly carcass.
I wanted

something to conjure with, not magnetic sudd. Even so,
I guess
they'll strip-mine it clean as a wiffle ball, leave it
derelict—
a drafty celestial rind. I just can't get over
so round
a monstrosity. Imagine something that big being dead.

IX WHEN YOU TAKE ME FROM THIS GOOD RICH SOIL

When you take me from this good rich soil
to slaughter in your heavenly shambles,
rattle my bone-house until the spirit breaks;

no heart of mine will scurry at your call
to lie blank as a slug in the ground where
my hips once rocked and my long legs willowed.

No heaven could please me as my lover
does, nor match the bonfire his incendiary eyes
spark from dead-coal through my body's cabin.

When, deep in the cathedral of my ribs,
love rings like a chant, I need no heaven.
Though you take me from this good rich soil,

where I grew like a spore in your wily heat,
rattle my bone-house until the spirit breaks;
my banquet senses are rowdy guests to keep,

and will not retire meekly with the host.
When, midwinter at the gorge, I saw
pigeons huddling like Cro-Magnon families,

no seraphic vision could have thrilled me more.
When you take me from this good rich soil,
and my heart tumbles like the chambers

of a gun to leave life's royal sweat
for your numb peace, I'll be dragging at Earth
with each cell's tiny ache, so you must
rattle my bone-house until the spirit breaks.

54

CAPE
CANAVERAL

CAPE CANAVERAL

Miles beyond the inlet
crazed with pelicans
and the puddyblue churning
of the Banana River,
an iron thatch
stood its ground
like a sentinel,
gripping the stiletto
rocket to its heart.

The sky put on a summer frock.
Cloudbanks piled
like a Creation scene,
dwarfing all
but a ghostly trail
embossed on the gravelpath
below (like the tread
of some rampaging mammoth)
where, earlier,
a tons-heavy slug

creaked its towering heft
along, to ferry
the Viking craft upright.

We glued our hopes
to that apricot whoosh
billowing across the launchpad
in spasms, like the rippling
quarters of a palamino,
and now outbleaching
the macaroon sun,
as a million pounds of thrust
paused
a moment
on a silver haunch,
and then the bedlam clouds let rip.

Gnats capered everywhere
in the marshland viewing site,
driven from their quiddities
by the clamor
pealing tindery to world's-end.
And how I envied
the wheat-colored moths
flitting about in a spry tizzy,
blind to that rising
persuasion called flight
groaning on a tower downwind.

I knew surf-jockies
rode their rollercoaster sea
on this ordinary day
in an ordinary August;
couples huddled on beachtowels
as if on loveseats.
Perhaps they'd see
an odd blaze far off,

Viking slide into the air
like a flint into water.

I was thinking of vigils:
radar hubs
following the craft
like sunflowers,
tracking dishes worldwide
now and again
rolling their heads
as if somehow
to relieve the tension,
how we'd gathered
on these Floridian bogs
to affirm the sanctity of Life
(no matter how or where
it happens), and be drawn,
like the obelisk we launch,
that much nearer the infinite,

when iron struts
blew over the launchpad
like newspaper,
and shock waves rolled out,
pounding, pounding
their giant fist.
My highflying pulse
dove headlong,
and then, like a cagebird
whose time was due,
my heart lifted off
into the breath-taking blue.

MARS

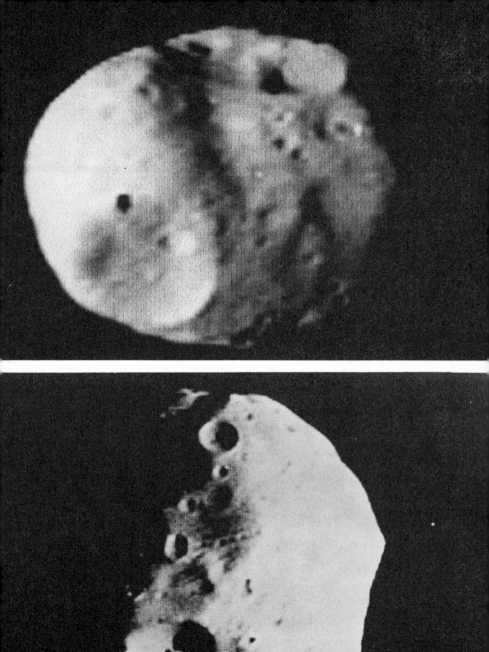

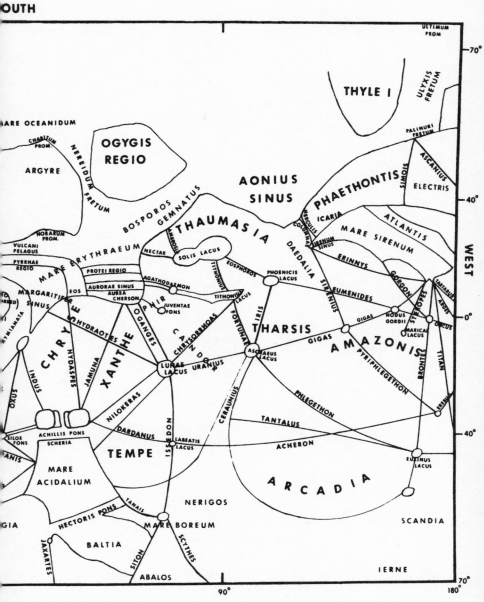

MARS

The quickest route
from *Candor* to *Chaos*
follows *Coprates*
(the much-travelled
Shit River), through
da Vinci and *Galileo*,
bypassing Bliss,
many moons from *Tranquility*.
But, Romantics, take heart:
you can breakfast
in *Syria*, lunch in *Sinai*,
track the Nile
to its source (*Nilokeras*)
before dinner, and there,
making ablutions to *Osiris*,
win a boon to *Eden*,
where all four rivers
of *Paradise* converge,
then spend the night
in *Pandora*, or with *Ulysses*,

Proteus, or even *Noah*,
in the Land of Gold (*Chryse*)
or by the Leek-green Sea.

That madman Schiaparelli
took one peek
at the Martian tabula rasa
and, daft with the sheer
profusion of matter,
went dub-wild,
leaving here a *Daedalus*,
and there a *Ganges*.
Now we can ankle off
to *Ultima Thule*,
hold the first annual
space games
on the Snows of Olympus,
found a divorce-mill
near *Lethes* (Forgetfulness),
and send anal-retentives
to *Coprates*.
I, for one, prefer
my psychodrama straight,
would pass over the equator
at *Cyclops*, ford
the Crossroads of *Charon*,
and follow the river *Styx*
to the Lake of *Hecate,*
the underworld queen.
Or elope to *da Vinci.*

Leonardo, you were the best
inventor: a colossus straddling
the old and the new,
buoyant and uplifting
as the vulture you saw
swoop down at your cradle,

air-strumming with a fringed wing.
Though your predilection
ruled women out, could you deny me
in *Uchronia*'s unhurried oasis,
or in *Eos*, that perpetual
rose-toed dawn? And what pigments
you would have made
from these Martian ochres!
Basalt, pumice, limonite,
obsidian . . . I've powdered them all
and mixed them in media,
drawn off umbers
that would make you blink.
You could lose a sorrel gelding
in them, or a roan.

As for Mercury and Venus,
with all their ignominious craters,
whatever will we call *them*?
I get twitchy not knowing.
I'm not asking the Universe
for a stay of execution,
or even that the Logos
declare itself, only
for a quirk or two
my curiosity snags on.
Right now, as I sit here
at the kitchen table,
I want someone to sashay up
to the door, and calmly say,
"Lilith, Ojibway, Rasputin."
I won't plague him
about quarks, cancer, UFO's.
I promise not to ask
about the spurred flagellum.
Only send word. You can trust me.
I listen for a knock, hear none.

Orange-ochre tonight,
ice-age Mars gutters
like a Turner sun.
Brilliant white polar caps
wax and wane with
the weathery (not the growing)
seasons. I know
the rough, bouldery terrain
and volcanic sputum
by heart:
think of the pumice soap,
the obsidian jewelry
(like folds of black satin),
limonite speckled brown
and mustard (on each grain
a fleck of iron:
its own tiny anchor),
the pocked lava rocks
that look like red sponges,
the charred basalt
grilled in mid-pour.

Perhaps Mars
fell together once
long ago, and now,
in an ice-age,
its former atmosphere
tucked discreetly
under a polar cap,
awaits the coming
of another spring.
Meanwhile the winds
chafe like emery boards,
carving rock into freeforms
and sway-backed arches.

There was a climate here
once, running water
and the blossom urge,
where sinuous dry riverbeds
stand out now
like veins on a temple.

Yet Olympus Mons,
the largest volcano
in the whole solar system,
may erupt tonight . . .
or not for a century.
I hope it will tonight.
I try to imagine
a mountain 20 miles high:
7 Alps perched
one upon the other's shoulders.

How can it be
this lily-livered flesh
houses the same atoms
that built the Andes
and Himalayas?
I am ashamed
for my lymph nodes:
paltry nubs, even
on the planetary scale.
My arms cascade,
knowing not where to go.
Each day, my mind pitches
its tent elsewhere.
And suddenly,
I have so many toes.

I think how land
forms: massive, bulky,
into stolid plateaux

and great brawny sierras,
and am startled
what touch-and-go
creatures we are
on this minor planet
of a humdrum star.

Wherever I look there's catastrophe:
          craters within craters, volcanic ridges,
                    giant impact basins
like old bull walruses
          gutted by plummeting debris,
                    collapsed lava tubes and braided channels,
sand-dune fields 300 feet tall,
          the rift valley in Coprates
                    nearly longer than all Africa.

The wind rips by at half the speed of sound,
          then turns tail
                    crossing the equator,
here and there
          scours an oval away
                    to show basalt gleaming like mealybugs.

In the Hellas basin
          (I prefer to call Hell's Kitchen),
                    a dust cauldron boils over
to storm-wreck the planet.
          Poor Phobos, the battered child of Mars,
                    looms overhead, gouged out and broken.

All around me:
          planet, moon, sun, riverbed, marsh:

grew out of cataclysms galore;
nothing ever sprang whole, stays put.
I feel the earth beneath my feet
suddenly shale away;
everywhere I look there's a new disaster,
and what splintered the mountains
made gape the pine.

A spelunker's dream,
Phobos and Deimos (Fear & Terror)
chug past, catacombed to the hilt
from all those years
of knocking about, and blacker
than the blackest pitchblende.

They look two prehistoric skulls
unearthed by a Leakey
in some intimate recess of Olduvai Gorge;
you can just make out
the eye-socket and jaw, the sunken cheek
and forehead spur.

From another view,
two porous kidneys hover,
long-suffering in the cellar-dark,
as if awaiting a transplant,
begun long ago, that will take yet
another eon to complete.

Till then, I challenge all comers
to the first game of Phoball (Earth rules).
No fear, you'll know my team on sight:
we'll be the ones vaulting skyward

like slow-motion gazelles, thinking how
more than our hearts leap up.

So this is where Elysium lies,
just north of Atlantis,
on the far side of Barsoom:
a velvet landfall, where volcanos
dome gently like saucers
overturned, crater-crowns blend,
no sharp conelike rises ever snag
or grow thorny, and canyons
meander like loose twine. And there,
delirious in our padded cell,
while two moons climb and reclimb
a purple sky, we'll see late-
afternoon clouds lurch overhead,
and the grinding-wheel wind buff
obsidian till it gems in the light.

And, when in summer the bald noon sun
dims behind a dust arroyo, let's
board a nightfreight to Lemuria
(string of opera-length craters)
and, like Laplanders, dot the polar
steppes, while electronic cows nibble
the hoarfrost, their udders ripe with water,
in a land where even ice sublimes.

Love, fly with me to Utopia:
three majestic snow-cowled volcanos
poking up through the sockeye dust.
Like Sherpas, astraddle our mechanical
goats, we'll guide parties
all across the chapped terrain,
early seacliffs and ochre pastures,
tending our rock-leeches that suck

mineral and water till, gorged,
they thud like gekkos to the ground.

Come away to the highlands
of Tharsis, and watch the red world
simmer below, teeming with dust-devils
and stiff black shadows,
towering sand-dunes, lava plugs.
Once in a blue sun, when volcanos
heave up grit regular as pearls,
and light runs riot, we'll watch
the sun go darker than the sky,
violet dust-tufts wheel on the horizon,
amber cloudbanks pile, and the whole
of color-crazed Mars ignite.
Come make a dun mare of a wind-carved
arch and, as the rusty sand blows past,
we'll dream ourselves a-gallop
this side of Tranquility, just beyond
Utopia, and through the Martian moors.

ASTEROIDS

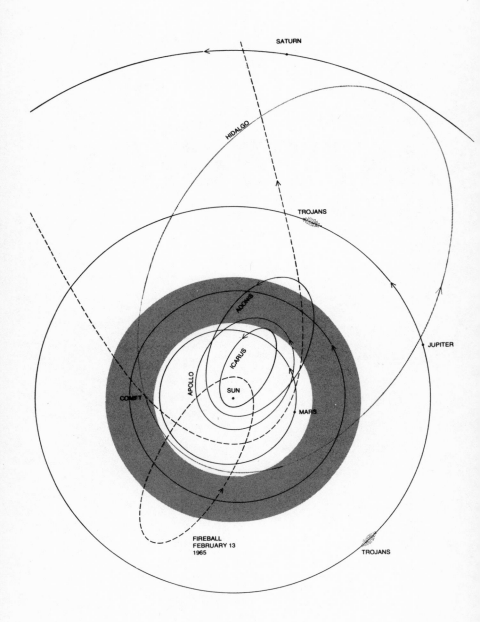

SATURN

HIDALGO

TROJANS

ADONIS

ICARUS

APOLLO

SUN

COMET

MARS

JUPITER

FIREBALL
FEBRUARY 13
1965

TROJANS

ASTEROIDS

We imagine them
                    flitting
                                    cheek to jowl,
            these driftrocks
                    of cosmic ash
thousandfold afloat
                            between Jupiter and Mars.
Frigga,
            Fanny,
                            Adelheid,
            Lacrimosa.
Names to conjure with,
                    Dakotan black hills,
        a light-opera
                            staged on a barrier reef.

And swarm they may have,
                    crumbly as blue-cheese,
            that ur-moment
                            when the solar system
broke wind.

                    But now
          they lumber
                                    so wide apart
          from each
                    to its neighbor's
pinprick-glow
                              slant millions
            and millions
            of watertight miles.
                              Only in the longest view
      do they graze
            like one herd
                    on a breathless tundra.

78

JUPITER

JUPITER

I

Vibrant as an African trade-bead with bone
chips in orbit round it, Jupiter floods the night's

black scullery, all those whirlpools and burbling
aerosols little changed since the solar-system began.

The mind reels to berth so gelatinous a rainbow,
suddenly pale salmon, then marbled blue.

Even
that lightning-
prickly red hurricane,
scouring the planet-face like
a pox, tones down to
sienna, deepens
to plum.

81

An agitated fossil, where no future is
no past, Jupiter lolls in its palatial rut.

II

Hydrogen, out of it the Universe evolved,
every atom and leaf, marine iguana

and apricot-smelling chanterelle. But my, my,
what alchemy: nondescript H_2

into voluptuous consciousness. I only wish
I could return the favor. Ripe in Jupiter's

wad, the chemical building-blocks lie low:
slipshod, aimless, on the brink.

But cellular puberty comes late. I mean
sex among the polymers is something

to ballyhoo. Carbon's atomic wildcard
must form one liaison after another, nimble

as a well-heeled paramour; amino acids
flourish, and primeval broth receive its

genetic *bouquet garni* before life starts
gunning for something creatural: a bacterium,

an oak, a slipper-shaped liver, or that ultimate
hand-me-down, Man, who minds the world

alien and overawed, a sac of unfelt gestures
and ungestured feelings.

Should life rinse Jupiter's paintbox bayou,
I swear I wouldn't pause a millisecond,

cheering now a pack of viral goons, and now
that molecular epidemic called a rose.

III

I sing the night sky. Ah, Capella, how you
flick your pollen heels when Jupiter mounts

the ridge of Aquarius, his lewd bits of fluff
all dancing attendance: Io, in whose transport

he hulks tight as alum; incontinent Callisto
twirling like a diva; Europa, the heart throb

who receives him as a bull; and Ganymede
(the one place *colonization* sounds apt)

for when he prefers altarboys to concubines.
Homer says the thunder-grey clouds of Zeus

burn tasseled with gold, and, aegis-center,
a howling Gorgon rears its head,

wound-mouth agape. When Zeus nods, lightning
splits the featherweight sky; when he rouses,

perdition and horror craze the earth.
The balletic moons pass, gala in his wake.

You talcy minxes, keep him sweet.

IV

Jupe miners, sifting the migrating ocean
for flocculi to power Earth plants and cities;

thermal-riding navies, tending bionts penned
in lowlevel seaquaria; lighthouse keepers who,

using Jupiter's heft, slingshot to outplanets
and beyond: off-duty, all these will roller-coaster

round their lunar circuit, consulting the moons
like a toss of bones. Perhaps children will school

at a satellite campus . . . on a *real* satellite till,
draft age, when they do alternate Earth-service

guarding prisoners exiled to Ionian salt-mines,
where not toil, but methane snow, foxes one.

On Io, binding the salty uprush, each flake
falls irregular, uncrystallized:

dud snow in a placebo season, conjuring early
winter in upstate New York—berries freezing

to hard-centered nuts, rhododendrons pointing
their digits sunward, bears eating tree-bark jerky

instead of grubs and freshwater bass. Zenith
to horizon, they'll see fields of color bloom,

and the far-cry Sun, no larger than a buglamp,
set quickly: night's tiny yoke sucked out.

V

The lowing of a neutron star
summons me to Jupiter's lanterny rheum

while his old besotted loves churn
and churn about: Callisto, Europa,

Io (the red sentry), Ganymede, *et al.,*
clear out to Hades, beyond which

other suns oblige other planets,
and gruel between the galaxies

pours down lightyears like a serum.
I wonder how far that is in fathoms.

Elsewhere, life homesteads
its many-chambered heart with theories

ill-timed and tentative.
Little humps appear and vanish and little

scuffles: other planetarians up early
doing their chores. Perhaps they haven't

two hopes to rub together. Does their moon
slide into the night's back pocket,

just full when it begins to wane, so that
all joy seems interim? Have they arts?

Do waves dash over their brains
like tide-rip along a rocky coast?

Are they flummoxed by that millpond, deep
within the atom, rippling out

to every star? Do they greet the Universe
with unfulfilled longing? Even if

their blood is quarried, I pray them well,
and hope my prayer their tonic.

SATURN

THE MOONS OF SATURN

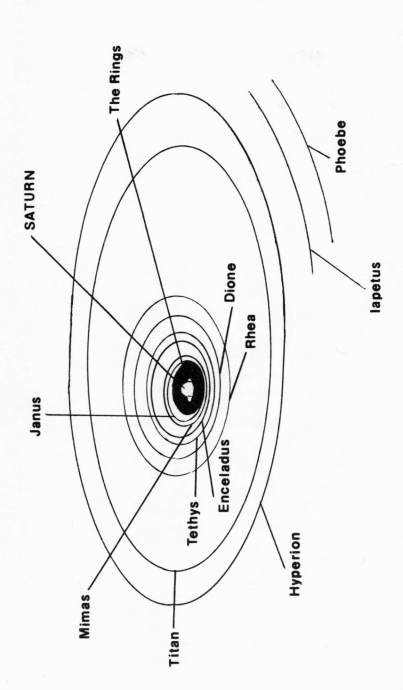

The Rings

Phoebe

SATURN

Dione

Rhea

Iapetus

Janus

Mimas

Tethys

Enceladus

Titan

Hyperion

SATURN

My sugardaddy bribes me
with a yellow-white brooch
(striped-enamel, diamond,
and ten glittery baguettes)
dangling in a black velvet box.
I stretch up for the honey-pale token, to pin it
on my blouse, over a heart lit like a jack-o-lantern.
But Saturn lumbers off
with its curio cloudbank,
sky-tethers the icy shoal
elsewhere, padding round
an orbit, just out of reach.

An elliptical blur creeps
into my field of vision, bellowing
light, as I wind the scope down
to such razory focus that Saturn
lies stunned in a hall of mirrors,

hog-tied by the cross-hairs, a little green at the pole.
Today lawnmowers studdered, Pekingeses yapped, trucks grumbled
        through downshift arpeggios; but tonight,
        risen in the botherless black,
        Saturn's ball of lemon ice
        looks so cool and reviving,
        I track it like the Golden Fleece.

        Millions of vest-pocket moons
        hang together as rings
        that loop round the planet
        like a highway skirting the golden city.
        Dusky bright, and godawful sheer,
they dog the equator (like Uranus's moons), never more than
two miles thick: a sprawling coral reef of tailless comets,

        grinding one another
        finer and finer, lolloping boulder
        to dusty mote as, eddying
        down through the crêpiest ring,
        they pour into a gassy draw.

        A tiny moon's constant nagging
        shoos the odd guest
        from Cassini's Division,
        an open airway between the rings.
        Phobos-sized snowballs tidy it up:
sentinel, roughshod, and devout as gestapos. Wide-open,
the rings hustle tons of light (more even
        than the planetmain), coming on
        so strong in the winter sky
        that Dione and Tethys
        pale away, like streetwise cats
        ambling into the night.

In a seaquarium big enough
Saturn would float! lighter
than rock, or water. I marvel
it even holds together:
hydrogen-clotted ice, frozen
methane and ammonia, all lathered to a gaudy slush,
like Jupiter a bit, only colder, which may be why
the weatherworks and the lazy
cloud-roll idle as they do,
and ammonia freezes out
as a yellow blizzard, snowing
deep into the planetball.

Saturn nods, an out-size natural
sponge, adrift in the galactic shallows;
but far beneath the haze, in a rocky core,
20 Earths could be bedded: stowaway
planets tucked neatly inside,
like sharks napping in an underwater cave. I see the heart
of an artichoke, I remember Goya's *Saturn Devouring a Son*.
We couldn't live here, I'm afraid,
will have to stud the moons
with our kiosks and rotundas, eyeing
from below that striped hammock
bellied across thin air.

A yearlong carnival, known as
Saturnalia, begins with ritual
carving of the rings, when mythic
figures hewn from snowballs—
thousands of Minotaurs, Gorgons,

and Dilemmas, Atlases and Leviathans—all swirl round
the Lord of Halcyon Days, making breezy sacrifice.
   For, like Japanese sand-drawings
   in an earlier epoch, moon-
   carvings are meant to erode, be
   dislimbed by the hobnob and bump
   of the rings, till not a rack remains.

   Saturnalia's a sport time, too,
   when Jovian moon-surfers take to Saturn
   (another leg of their endless cavort),
   flying wildly round the rings' undulating
   carpet, or croquet-thwacking debris
in Cassini's Division (always in danger of being thwacked themselves,
and hauled away by a moonlet-tugboat). Anytime, you'll see the usual
   hawkers and parades, craftspeople, con-men,
   homely souvenirs; but only early season,
   before the crowds descend, can you watch
   the ice-masons hard at work, or hear
   the joyous *hoopla*'s of the ring-riders.

   Entry, *Fodor's Guide to Saturn:*
   "Best camera shots
   from Iapetus or Phoebe. . . .
   Avoid Titan (too cloudy).
   The other sherbety moons,
all smackdab in the ringplane, make Saturn appear
utterly ringless:
   an agate bulb, with a tally line
   summing mid-planet.
   But, viewed from Iapetus,
   Saturn swivels like a gyroscope,
   its hatrim turned

93

up and down,
while the planetcore stands still.
Daytime, you see back
and darkside of the rings;
nighttime, the sunlit maw.
Only be sure to book a yurt on Planetview (the side facing
Saturn); on the other, glued always to deep space,
you could live out
a lifetime, never knowing
behind you lay
a lighthearted planet,
maizey in a halo of ice."

On Titan, warmed by a hydrogen blanket,
ice-ribbed volcanos jet ammonia
dredged out of a glacial heart. Liquid
and frozen assets uphold an empire
bigger than Mercury, and even a little
like primitive Earth: asphalt plains and hot mineral ponds. But
how I'd like to take the waters of Titan, under that fume-ridden sky,
where the land's blurred by cherry mist
and high above, like floating wombs,
clouds
tower and swarm, raining down primeval
bisque, while life waits in the wings.

Often I dwell on the Big Bang,
find my heart levied high, and the vision electric,
am wowed by that arch creativity.
When I tell people, they flinch
with terror, want no part of the ur-inferno,

will not truck with apocalypse. But Paul at the scope, one finger
on the clockdrive, tunes in the Universe with the affectionate
curiosity of a naturalist.
And I know, if I trigger the mental
clockdrive, his mind will gingerly
backtrack and zoom, run rings round
the spectral notion of Saturn.

I say, *"After the never-ending*
gas cloud coalesced, the Universe
was all in one place,
and solid: a hard, local object
in an endless ether." He smiles,
says, "Wonderful plot!" *"In the beginning was the Word,*
and the Word was a tough, silky ball of hydrogen."
He splits the double star, Albireo,
then pulls back a moment,
says, "Just imagine the commotion
of the Big Bang!" We huddle
in the breath-taking dark, and imagine.

Tonight, what with the moon
keeping so low a profile, the stars
are bright as campfires. Waltzed
around by how many planets? Drenched
in how many groundswells of life?
My saturnine ring-leader, pallid-footed, strolls along
with ten swanky moons in tow. And though I'm smitten now
with this giant manticore,
heartwise I know it's only
a panaching fancy: somewhere else
in the odds-on of space, evolution
may be minting a pipefish.

URANUS

URANUS

(Scene: Pataphysics Symposium aboard the spaceship Euchronia, en route to Uranus. Over the loud-speaker comes a fluent lugubriousness; it sounds like spiders knitting, but is probably Richard Strauss's Metamorphosen. Sir William Herschel, Sir Isaac Newton, and four Astronomers Royal—John Flamsteed, George Airy, Edmond Halley, and William Christie—are seated at a long banquet table on the far side of a vaulted chamber, dining on cornish hen and Jello molds resembling each of the planets. Heatedly, they discuss what Uranus may look like, and other pertinent matters, seemingly undisturbed by the witches' sabbath unfolding in the middle of the room. Herschel's sister, Caroline, leads the band of witches—Annie Jump Cannon, Angelina Stickney and others—all cackling and hooting, and riding naked on shaggy goats with twisted horns, on telescopes, bathbrooms, pokers, and grunting boars, casting enormous shadows as they dance around a fiery black throne where the Unknown sits in state, surrounded by horror. Above the festivities, a vapory armada of tiny love-junks hovers at full sail. Caroline hands round a pot of greasy green unguent.)

Caroline Herschel: Lizard piss, claws, shaggy spiders' paws,
lethal lactuca, fat of unchristened babe,
eyeball, batsputum, seepy white wart,
Hyrcus Nocturnus, come to our aid!

Witches: *Hyrcus Nocturnus*—come to our aid!

Caroline Herschel: Great Goat of the Night, transport us a far
passel of parsecs from this local star!
Swamp smallage, hemlock, honey-bee drone,
come, little sisters, we'll guide him home
on bagpipes made of dead men's bones!

(*Squawking and howling, they canter around the fire, which jumps to the croaking song of the bag-pipes.*)

William Herschel: ... I say it's green as avocado or unripe pomegranate,
lying sideways, and askew *thus*, like no other planet,
with one pole where an equator should be,
its gaze tilted sunward, as if on bended knee.

George Airy: So each day lingers on through some forty-two hearings?
The bilious midnight-sun must be searing!
And another effect, about which we know nothing,
say, a photon as addled as a cage-weary gosling,
on the whole, could render one down in a whistle
to a saucer of spunk, or a retort of gristle.

Isaac Newton: Gad, how you fable! You're both full of Scheat
you know: stars and not planets. What nonsense about heat!
And debate makes me ill; could you manage less row?
I've only just unpinioned this fowl.
Uranus, anyway, is glacial. Measurements show it:
minus three-hundred degrees Fahrenheit (even below it).

George Airy: Nonetheless, there are many who'd say
at the pole, all the heat of the sun makes the day
like a hornet, while elsewhere it's perfectly icy.
Come now, admit it, *your* theory's as dicy.

Isaac Newton: *I* admit nothing, am trying to eat
this bird before you all put me to sleep.

101

Halley, I note, is already out cold,
bleating like a lamb dragged off from the fold.
And, as for Uranus—by Jove, what a pest!—
look, if it's toppled, as you all protest,
then the caterwaul winds blow east (not west).

William Herschel: At last, on something we concur!
(*Aside*) I do wish someone would give Halley a stir.

John Flamsteed: Why? He so often says more than he knows,
who cares if he's in crypto-biotic repose?
Oh, all right. Come on, Ed, join the living.
(*Flamsteed elbows Halley, who wakes*)

Edmond Halley: Grmmmph. I think an ox is standing on my tongue.

John Flamsteed: That's Aeschylus. Honestly, he's like a chafing-dish.

Edmond Halley: What words escape the barrier reef of your lips!

John Flamsteed: What, Aeschylus *again?* Christ, you were sent here to pain us.

Edmond Halley: And you, sir, have a mouth like a starfish's anus!

Isaac Newton: Gentlemen, gentlemen, let's hear the end of it.

William Herschel: Yes, grapple with Uranus, what we comprehend of it.

Isaac Newton: Bury the axe; your feud's getting woolly.
Wounds heal (except those we inflict, hopefully).

(Christie clinks his fork against his water glass, and rises)

William Christie: Dividing my talents, I have here a brief
diversion to offer for comic relief:

(He reads perkily)

"Amelia Earhart, the goddess of flight,
crossed the Atlantic in the dead of the night.
She broke the air records and the hearts of her fellas,
who never could get her to twirl their propellers!"

Isaac Newton: Droll, very droll.

John Flamsteed: Droll? It's disgusting. And him a friend of Edward VII.

Edmond Halley: *That's* not worth the spleen of a Portuguese kipper;
Edward VII was Jack the Ripper.

John Flamsteed: Harmless he may be in the observatory,
but elsewhere he's downright predatory.
Through a telescope, he gapes so snake-eyed
even decorous Venus averts her backside!
And I personally foiled his attempt to vend a
Trifid Nebula arrayed in gynecological splendor,
the Seven Sisters posed lewdly with the Lion,
not to mention the "golden nuts of Orion,"
and his hint that black holes. . . . .

William Herschel: Flamsteed, really, this is quite out of hand.
Our far-famed colleague is an upright man.
Besides, we're lightyears away from our plan.

Isaac Newton: Yes, indeed (I recommend the peach-flan).

William Herschel: Let's press on now, and give the moons a whack,
then tend our drier business over cognac.
Perhaps Christie, a peerless ruminator,
would suggest why the moons so hug the equator.

William Christie: I imagine some meteor, swaggering like a trollop,
bumped into proto-Uranus, knocking a dollop
of vapor away from the planet's middle,
which later condensed as Uranus twiddled
five gassy fingers into moons
quite unlike Saturn's billowy cocoons.

William Herschel: But I think you'll agree
the moons could as well be debris
from the collision.

William Christie: It's possible. However, I've an ambition
to tuck something into Euchronia's record

before we proceed on our aero-buckboard.
As all the moons—Miranda, Umbriel,
Titania, Oberon, and Ariel—
are named after fairies & sprites in a scene
of Shakespeare's *A Midsummer Night's Dream*,
let's agree if another moon turns up
we'll keep to the pattern and christen it "Puck."

John Flamsteed: Bravo! So motioned.

William Herschel: And seconded.
Good gentlemen, just for the sake of clarity,
Christie raised a point about which there's disparity.
When first I eyed Uranus, it was dragonfly green.
Of course, later, faint grey belts were routine,
and cloudbanks of methane dark as fern,
but I'm sure it looks nothing like old Jupe or Saturn.
Why, it's nowhere as big for a start,
and that means it's nowhere as hot.

(to Newton, growing ill)

106

George Airy: On my word, you look green as old bouillabaisse stock.

Isaac Newton: I've a hunch it may be anapestic shock.

(The witches break into a loud chant)

Witches: Hebe, Recha, Bilkis, Lacrimosa,
Eucharis, Betelgeuse, Grubba, Ambrosia,
Eunike, Orchis, Fanatica, and Frigga,
Nymphe, Nemesis, Hippo, Walpurga!

(A bat flutters through the banquet hall)

John Flamsteed: RU Lupi? Hot or not? A specious dilemma.
Even lying on its side, as if . . . as if . . . .

William Christie: Waiting for an enema?

John Flamsteed: *(whispering to Airy)* Do you know he plied on a Zambesi stutterer
"the man on the flying Trapezium" and Mas-tab-ba-tur-tur,
along with a fish-eye photo of "Lick" . . . .

William Christie: You forgot the exhibitionist
Sun, stepping behind a cloud to open his raincoat.

George Airy: Oh, I *get* it! I do, and I like the joke,
am surprised you could miss the coal-sack in the Swan.
How about the red giant, an Alde Baron,
and especially my own gosling photons,
or the Argonaut Ship, that *rara avis*
doomed to eternally stare at its Navis!

William Herschel: Now, gentlemen, once again I implore you.
Flamsteed, I believe you had the floor hitherto.

John Flamsteed: Look here, I was saying: with no internal heat-gun
(I mean dynamo, like Earth, Jupiter, Sun),
Uranus must have an atmosphere stable
as a prop-moon, or a backdrop of sable,
despite some local windy regions.
More clement than Earth. Just think of the queer seasons!

Isaac Newton: Gibberish! Hell, where do you get a season,

when the Sun only gives the *pole* its secretion?
Furthermore, all you've said is plain iffy.
So why bother? We'll be there in a jiffy.
Fuel up now. Later you'll sample its edifice;
just keep an eye out (though not like Oedipus).

William Herschel: Perhaps this *is* high time to pause;
our agenda's developed so many flaws.
And, too, I've been handed this woeful note:
it appears poor Barnard dropped dead at his scope.

John Flamsteed: My, my.

William Herschel: Come morning, they found him curled up like a prawn.

George Airy: Well, at least he died with his Boötes on.
(Newton, eating an enormous bowl of fruit, sprawls across the table,
like a boa constrictor digesting a zeppelin)

Isaac Newton: Yum! These apricots from Fomalhaut are like a creed,
there's so much meat to protect the seed.
Juicy and ripe as a young consenter. . . .
(Oh, dear . . . I'm eating the fruit's placenta!)

(Herschel opens a thick dusty tome, and riffles through the first 10 or 12 pages, drawing one out)

William Herschel: It's time now, alas, to deal with prosaics,
and today's calls for Penelopean finesse.
Winnow, we must all those who apply
to our august society, the others dis-ally.
First, what about Nathanial Bliss?
To neglect him, I think, would be remiss.

Edmond Halley: As astronomer,
he's anonymous.

John Flamsteed: His sole portrait adorned a mug of taproom-piss,
below which it read, "On earth, this is Bliss."

William Herschel: I hear he had cataracts and a wife
you could slice bread with.

(a hasty vote carries)

Next on the docket . . . this will make you rant:
D.S. Ackerman again suing for a grant,

her letter phrased, I'd say, quite ruinously.
But you decide; the application follows thusly:

"If one begins merely to end, then
I have only just begun to stop. . . ."

Isaac Newton: Promising,
but do edit the thing.

William Herschel: "I am requesting funds to study what extra-terrestrial creatures would need in order to write poetry. For example, would they need internal pulsing organs, or get around in iambs, or have electric brains which facilitate visual puns, *viz.* metaphors. . . ."

Edmond Halley: Wasn't she the one . . . not that I blame her. . . .

William Herschel: As I recall. But this year she includes a disclaimer:

"I am no longer responsible for any acts committed by my previous selves."

George Airy: You know, actually, that makes sense.
Read her answer to our query about "poetic license."

William Herschel: She says, she only just earned her learner's permit, and "poetic licentiousness seems more requisite." She says,

"it's fall again: the leaves like red tongues; death putting up a brave front"... needs "to truss up some vague prolapse of the spirit on nothing less than Orion's belt...."

John Flamsteed: Oh bother! The world's storied with such laments. I say we've given her too many grants already. Last time, it was "a hot vapory mid-day in July," and before that some other season gone awry.

William Herschel: We can at least hear out her references; the Universe could use a new amanuensis.

(*Herschel hands round a sheaf of papers*)

Do note, Paul West says:

"Miss Ackerman is like a goldfish that swims to the surface with an aphorism in its mouth."

Isaac Newton: Catchy.
                But the rest of her folder looks patchy.
                Nor do I care for "experimental" sobs;
                her traditional roots show like a bad dye-job.

John Flamsteed: She's utterly surdic. Brittle as sandalwood.
                How can such virtuosity be any good?

Edmond Halley: She ties herself in more knots than a hagfish.

John Flamsteed: You ought to know.

Edmond Halley: Bog off! One more allegation made against me
                and I intend to find out who the alligator is!

John Flamsteed: Up Uranus!

Edmond Halley: Ophiuchus!

John Flamsteed: Cosmic bourgeois!

113

William Herschel: As this is *not* the Baden-Baden Spa,
I suggest you two save your glut of ill humors.
Focussed, it could shrink a dozen tumors!
While you tinderboxes were banefully bickering,
I received a regal letter from W.H. Pickering,
who reports that one day last June
he spotted a flock of insects on the moon
living in "platts," and (what a surprise!)
traipsing about at every sunrise.

Isaac Newton: Poppycock. Clearly the man is besotted.
We should kick him out (then have him garotted).

*(In the middle of the room, the Unknown begins to fire
and glow like a pool of burning ambergris)*

Witches: Algol, Blinking Demon, your lurid factotum
we conjure on this pile of tortured cat-scrotum.
First, bring us Sirius, the great blue-assed satyr,
with his dark companion of degenerate matter;
then the Hyades' Moist Daughters in Aquila's grip;

114

Capricornus, monster half goat, half fish;
Zubeneschamali, riding Libra like a sow,
crying Oh Be A Fine Girl Kiss Me Now!
We writhe like flames over enigma's bones.
Upend! Reveal! Lay bare! the Unknown.

William Christie: A thought on ending (though nobody asked):
let us perform a Uranian masque!

(*Christie leaps up, thrusting his arms out stiffly, and begins spinning around the rim of the table*)

Ike, you can be Umbriel; Bill, Miranda;
Ed, Ariel; George, Titania. . . .

(*others follow*)

I'm Oberon: the kingpin circulator.
Step lively! Remember, we're near the equator.

115

(Cheered on by Christie, the astronomers twirl around the banquet table, each one describing a Uranian moon. Strauss's Metamorphosen gives way to a lyrical piece for electronic synthesizer, full of tinkling temple-bells and goblins farting. A greenish-blue Uranus grows larger through the porthole, while the Unknown, the color of ginger-root and henna, first glistens and then begins to dim. Christie unfurls a large phallic diagram entitled "Anatomy of the Magnetosphere," its parts neatly charted as a side of beef, from Plasmapause and Polar Cusps to Bow Shock and Magnetic Tail. As the music's ear-rending staccato mounts, the astronomers reclaim their seats, falling to the last of their topics: If Uranus will be solid or liquid. Finally, all one can hear of the Symposium is the occasional cadence of wilted mumbling; and even that, as the fires dwindle, fades.)

Christie's Diagram

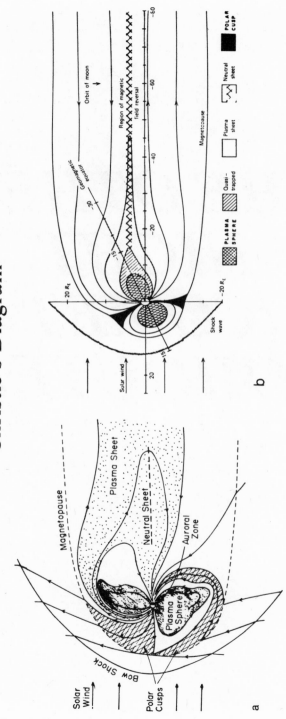

a

b

Solar Wind

Bow Shock

Polar Cusps

Magnetopause

Plasma Sheet

Neutral Sheet

Auroral Zone

Plasma Sphere

PLASMA SPHERE

Quasi-trapped

Plasma sheet

Neutral sheet

POLAR CUSP

Geomagnetic equator

Orbit of moon

Region of magnetic field reversal

Magnetopause

Solar wind

Shock wave

20 R_E

−20 R_E

THE OTHER NIGHT (COMET KOHOUTEK)

THE OTHER NIGHT
(COMET KOHOUTEK)

Last night, while
cabbage stuffed with
brown-sugar, meat and
raisins was baking in the
oven, and my potted holly,
dying leafmeal from red-spider,
basked in its antidote malathion,
I stepped outside to watch Kohoutek
passing its dromedary core through the
eye of a galaxy. But only found a white
blur cat-napping under Venus: gauzy, dis-
solute, and bobtailed as a Manx.

Pent-up in that endless coliseum of stars,
the moon was fuller than any Protestant
had a right to be. And I said: Moon,
if you've got any pull up there, bring me
a sun-grazing comet, its long hair swept
back by the solar wind, in its mouth a dollop

of primordial sputum. A dozing iceberg,
in whose coma ur-elements collide. Bring me
a mojo that's both relict and reliquary.
Give me a thrill from that petrified seed.

Mars was a stoplight in the north sky,
the only real meat on the night's black
bones. And I said: Mars, why be parsimonious?
You've got a million tricks stashed
in your orbital backhills: chicory suns
bobbing in viridian lagoons; quasars dwindling
near the speed of light; pinwheel, dumbbell,
and impacted galaxies; epileptic nuclei
a mile long; vampiric moons; dicotyledon suns;
whorling dustbowls of umbilical snow; milky ways
that, on the slant, look like freshly fed pythons.

Sirius began to stammer overhead, jittery
as a blue-assed fly. And I said:
All right, then, let's put our cards
on the table. I've never quite believed
that rumor of cohesive unity, in which all
things participate and adhere. Never felt part
of that black longwinded millennium
which, I'm told, is expanding with unhindered
clarity. Never could accept the extortionist
behavior of germs as part of some ebullient
inhuman will whose codicil's unknown.
And in this fleshy vellum, which I assume
to be mortal (there being no news flash
to the contrary), my mind can't conceive of
what it can't conceive of. I've got a brain
all dressed up and no place to go. Why don't you
hatch me a dream from that frogspawn cloud
of comets ambling up and down the carborundum
sky, whose sudsy white nougats of dust and ice
may be the only spirits in a lidless night.

NEPTUNE

NEPTUNE

I

Trident-slung potentate of the sea, mythic
Neptune twitches a bloodbelt finger over
water
and
a hurricane begins; he dials his anemone
hand one notch and fields unpelt, mountains
churn,
cobalt
and dolphin tangle like sea-brit, orcs toady,
nereids kowtow. Calm, above all, he's not
calm
like
the air leadening hot muggy noons in July,
but calm as the unsqualling lemon-hooded
sea
where
glory & terror nuzzle one another, itching

to combine. His children all were savage
harridans
and
giants till, one day, from his congress
with Medusa, hope-winged Pegasus was born.

II

Speckled Volutes squatting like tiny lepers,
dunce-capped Wentletraps, Carrier Bells,
Deer
Cowry
ballooned to geodesic domes, Turnip Whelks
spiraling like Edwardian staircases, comb-
                                    toothed
                                    Paper
Nautiluses, Jewel Boxes & Junonias, Triton's
Trumpets scale-modeling Mayan ziggurats,
Flame
Augers
turreted as Togoan mud forts, mandibled Apple
and Cambrit Murexes, Great White Spindles
                                byzantinely
                                engorged:

Salacia, Neptune's salty lover, rules them all.

And these but a caucus of deepwater shells.
Billions
more.
Then fish, plankton, mammal, coelenterate,
mineral, bacteria, mollusk, snake.

126

                                        Her
                                      clout
migrates with the salmon, diamonds the fossil,
insinuates the reptilian land. Her empire
unravels
from
the human body, salty too and yearning for
osmosis, out to the star-lapping cosmic foam.
                                       Sea-
                                     queen
and sea, tight-lipped and runic, she knows all
about the rigors of mortis: her own womb's
tureen
served
up the primordial soup.

III

Passing by,
I overheard you and Salacia in flagrante.
                                        You
                                      said:
"I'm afraid you're tight as a fist. I guess
my giant slalom's not up to par."
She
said:
"You've got more balls than Horace. Coyness
interruptus, on the hoof, or ad hoc."
                                        You
                                      said:
"But am I smoother going down than the ebony-
spleenwort?"
She

said:
"Adder's-tongue or stag-horn. Honey, you're
in deeper than a quagga in a quagmire."
                                        You
                                        said:
"My sibylline flirt, my cul-de-sac. Your
jellied womb froths like an avocado shake."
She
said:
"Your gonads are clean as two spring potatoes,
your synapses hard as three-penny nails."
                                        You
                                        said:
"My dear, you know my every passage by heart."
She
said:
"I should. I've been reviewing them for years.
Let me play you a tune for wail & skin-flute."
                                        You
                                        said,
agasp: "Literature becomes you."
She
said: "I barely mind."
                                        You
                                        said:
"Why don't I nibble your talced belly
as though it were a camembert?"
She
said:´
"Agggh, what a cunning linguist you are."
                                        You
                                        said:
"How ya' doin'? Or should I refresh your mammary?"
She
said:
"I'm fluid as an egg now and ready to be tupped.
Would you prefer me over easy or sunnyside up?"

Your voice pebbled, brooklike.
Hers gabbled and flew, now puckering like an almuce,
then clinging like a spore.

Snivelling and demented, I shook
till my unquarried metal fret apart and thick red
jams within me oozed,

then stole away, wanting
my own daylight Sibyl till I was low-down
and rabid as a skink.

IV

Fifteen billion years ago, when the Universe
let rip and, in disciplined panic,
Creation
spewed
mazy star-treacle and resin,
shrinking balls of debut fire smoldered
and
                            glitched.
Revolving tantrums tore themselves in,
kernel-tight, then cooler, began shooting off
their
planets
like a brace of dandelions gone to seed.
Even stars don't live forever; old age makes
                      the outer
                        region
redden and swell (Betelgeuse's radiant flush
in Orion is as much a death-spurt as the whale's

flower
of blood).
Like the body, a star grows larger as it runs
down. Neptune once flecked a coral sun
                                    (yellow
                            now
as a scrub-diamond) that will puff beet-red
in five billion years, and suck the planets
back
to
where they began.

It's this business of mortality: rallying to
one trickle after another, like a tuber root-
bound
in
a time of drought. I know it's wrong to care
so for worldly things: lanky poplars swaying
                                    like
                        keening
women, pond-gasses mulling over soil & weed,
the lime-green torches on the candle-pine.
I
should
inquire only what is durable, lingering; be
zested, not by things that pass, but by what
                                    will
                            come
to pass. I know I *should*. But I do not think
I could.

V

You were not summer
            raisining the grape.

It may be that in your voice
            I heard crocuses heaving
                        up upon their roots,

but you were mortal as a junebug:
you crapped and wooed,
            were fallible, had allergies.

I knew you were not April
            spurning the ice-hipped weed.

It may be through your retinas
            I saw snapdragons butter-flock
                        like popcorn,
but I knew you were not life's
            vesper in the cell.

It may be when you were gone
            it seemed the gypsy moth unshrouded,
                        wysteria threned,

but you were mortal as a whistle
down the wind,
            a spark bouquet.

I knew you were not stasis
            bedded in the marl.

And though you were not fall
            silkening the corn,

> how I loved you
> > in the open heartlands
> > of New York.

VI

Aqua-blue colossus lurking in the outback,
Neptune wheeled round its solar bevel once
every
160-odd
years: aloof, unheard of, and ill-defined:
a verdigris stopgap in the cosmic pause
                                        not
                                            even
guessed-at for two millennia, then
wrenched from Uranus's mathematical rib.

Even now, as I write this unwitting archive
(misconstruable as exploratory verse)
                                    Neptune
                                        is
elusive as a dappled horse in fog. Pulpy?
Belted? Vapory? Frost-bitten? What we know
wouldn't
fill
a lemur's fist. Yet people chide me for
esotericism. I tell them: No, it's
                            obsolescence
                                        I
should fear. In a century, this poetic
desert noesis I call *The Planets* will be
nothing
more

than a cerebral eddy: 9 grains chafing down
a harvest silo: while Neptune's blue

> monastery
> houses

Benedictine monks tincturing bath salts, or
Simon of the desert atop a hoarfrost pillar,
or
a brambly
cloister in the planetary boondocks, where
pilgrims en route to Orion lay over and *The*

> *Canterbury*
> *Tales*

brew again. Perhaps even bog-burial
will flourish, as space-outlaws bent on
skull-
duggery
hide out, planting droves of freshly
garotted corpses in the shallower fathoms of

> methane
> and ammonia.

PLUTO

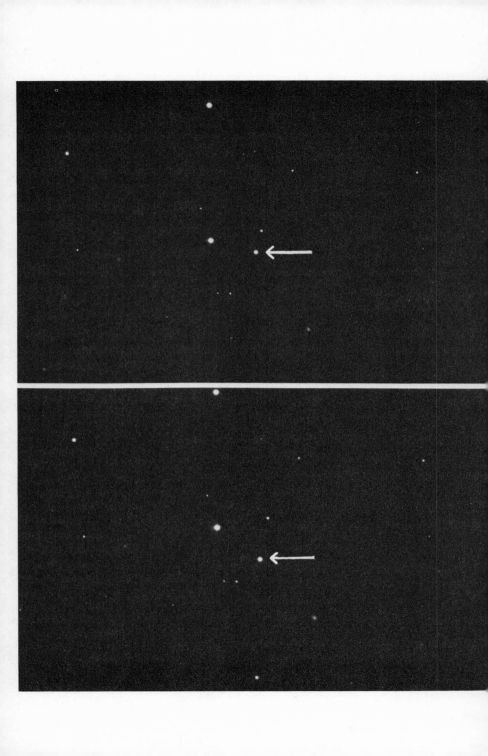

PLUTO

In my 26th year
I left the planet that bore me.
When the Sun had risen
like a golden fish
leaping high into the briny blue,
I put Earth behind me
and travelled light,
sailing out of my flesh
on the first good wind
and barrelling tide, to pace out
my tether in the hub of the Sun.
Those whom the darts
of wonder never fret
may think it odd
that on a vapory midday in July
a young woman
might take to the stars.
To these poorer souls,
how can I explain

what their own hearts refuse?
My need to know yammers
like a wild thing in its den.
I see above me Andromeda
in whose black bosom
galaxies swirl like pastry,
and I am so hungry.

At night I lie awake
in the ruthless Unspoken,
knowing that planets
come to life, bloom,
and die away,
like day-lilies opening
one after another
in every nook and cranny
of the Universe,
but I will never see them,
never hear the grumbling
swoon of organ pipes
turning the Martian high winds
into music, never ford
a single interplanetary sea,
never visit the curdling suns
of Orion, even if I plead
with all the fever of a cypress
tilting its spindle limbs
to con summer, piecemeal,
out of early May.

Once, for a year,
I was of nine minds.
And if I lacked nerve
somehow, clinging
to every image as if,
sandbags thrown over

and its balloon out lightweeks
on a flimsy thread,
my life itself might float away,
I knew the trail blazed out
was the way home, too.
How often my teeterboard hips
were desperate to balance
like a schooner's clock.
If I left with the fear of Antaeus,
it was not without
the faith of Eratosthenes,
who dreamt the world round
in a square age.

But now, 9 worlds later,
I hug the coastline
of yet another frontier: Pluto,
a planet conjured into being
by the raucous math
of Percival Lowell,
a land bristling with ice,
grey and barren,
where the Sun, nearly doused,
rallies but a paltry sliver
of light, and messages take
10 to 12 hours to field
(imagine the cool, deliberate
chessgames, the anxious lovers,
the crises exploding
between communiqués).
A planet-sized enigma
jogging in place, Pluto's moved
little since its discovery,
touring the Sun once
every 248 years.
You could be born in winter,
and never live to spring.

We think of Pluto as an endstop,
or skidding out
like the last skater on a whip,
a land glacial, remote, calm
and phlegmatic.
But right now, while you read
these lines (I swear),
due to an odd perturbation
in their orbits, Neptune and Pluto
are swapping places
in a celestial *pas de deux*
where the only aerials
are quantum leaps.

If Pluto has a menagerie
of moons, we don't see them
(nor, for that matter, the damned
wallowing in their slime,
or Cocytus, the frozen lake of Hades).
No, the Underworld God
keeps his dread secret a moment longer.
About Pluto, we've only
the odd hunch and inkling:
theories pale
as the wings of a linnet.

When our vagabond skiffs
breech the outplanets, I wonder
will we have the presence of mind
to call Pluto's main city *Dis*
(the hellish capital
Dante spoke of), or name
the ferryboat shuttle *Charon*,
the deadspace it cleaves the *Styx*.
Perhaps not.
An ocean is an ocean
after all, whether it loom

from Triton to Pluto
or Southampton to Plymouth.

Where are the Balboas
and the Amerigo Vespuccis
of tomorrow,
hot on the heels of the future,
who will give their names
freely, as if to wives,
as they voyage the spaceblack
waters, always going on
with restless ongoing,
to the end perplexed
by the force that sped them,
and leaving only their names behind?

If Pluto anchors
beyond our sweep, docking
far out along the midnight wharf,
we'll braise
our frontier towns on Triton.
How eerie its floe-broken lands
will seem, with no pink and green
wispy trees of summer
or every so often a blinding white birch.
Could I face only the galaxies
coiled like cobras?

Surely frontier towns
there will always be,
even if "town" seems
too fixed, too stolid,
for anything so mercurial
as "frontier" to be caught with.
Deep in the mountainsides,
where the temperature
is least likely to skitter,

we'll build
our snuggeries and hives,
be cave dwellers again.
It's as if, flummoxed
by the shock of living,
we step by step re-stage it,
driven to the most far-reaching
ritual. Like a catechism,
we begin again: the cave dweller,
the trapper, the trader,
the explorer; the self-reliance,
the hope, the patience,
the invention: wrapped in our past
as we breathe down
the glistening neck of the future.

Forgive my brain
its wanton poaching
on an earlier estate,
but such frontier talk
leads me back to da Vinci.

Leonardo, come steal
into the chamber of my thought
again. How I miss
that nomadic mind of yours,
always at red-alert
and surging like a furnace.
Often I dream that,
like a horse flinching
to keelhaul a fly,
I might shed the centuries
and give you a motor
or a fixed wing.
Had you erred into my bed
and body last night,
I could not hold you dearer.

What sort of woman can it be
who feels at home
in all the Universe,
and yet nowhere on Earth,
who loves equally
what's living and ash?
I can't seem to overlook
the context
in which I live,
this collection of processes
I call my life,
even though the flower
be indifferent to my pleasure,
and the honeybee virus
dragging its genetic pollen
from one cell to another
be blind to my despotic ways.
One sultry morning
I found a sneaker-print
in the mud, whose herringbone grid
looked like a trilobite fossil.
How you would marvel
at the alchemy of line.
All day I suffered
that I couldn't tell you.

The breadmold and I
have much in common.
We're both alive.
The wardrobe of our cells
is identical. We speak
the same genetic code.
The death of a star
gave each of us life.
But imagine

143

a brandspanking new
biology. Just as
when a window
abruptly flies open
the room grows airy
and floods with light,
so awakening to
an alien lifeform
will transfigure
how we think of ourselves
and our lives.
In my bony wrist alone,
the DNA could spin a yarn
filling thousands and thousands
of library volumes.
But one day we'll browse
in the stacks of other galaxies.
Given the sweet generosity of time
that permits the bluegreen algae
and the polar bear,
the cosmic flannel
must be puckered with life.
My bad habits charm me now
with reckless appeal;
we may be the habit of the Universe.

Today in the locker-room,
under the dryer,
threshing my long hair
in a wind that might have swept
off the Gobi or Kalahari,
I let my thought freefall
from Hercules, into whose arms
our Sun is rushing,
to the sky thick with planets
and ghostly neutrinos,

how through a telescope
color-flocked nebulae
look like cameos:
black and white miniatures
of themselves.

While such visions
and ripe polychotomies
waylaid me,
fleshed-up women
paraded by, whose breasts
swung like pendulums
chiming their hours,
and tummy-rises blurred
to an iridescent ripple.
Somewhere
far down the locker-row
a woman's voice,
like an eagle or kite,
balanced on a rising column of air.
I stepped out onto the beach
of our galaxy
and, as my hair became a trellis
in the solar wind, I wedded
that shining carapace of the future.

Once, for a year, my thoughts
gathered like clouds
into skycoves and jetties.
Entombed in a so-so body
coloring, I was perfectly wowed
by the Joseph-coat planets,
the lurid gas ribbons
and sherbety pastels,
Jupiter's organic chowder,
the saturnine rings bleeding light.

I consulted the moon
like a crystal ball.
I boned up on the flinty
inner planets, whose craters
do-se-do for miles.
I steered by Sirius,
the effervescent guide.
I pored over our bio-heirlooms
like a medium
needing to feel the murderer's glove.
I winnowed, I delved,
I compassed, I schooled,
breezing from one delicate
science to the next
with the high-flying rapture
of a bird of prey.
My heart jingled,
full of its loose change.

I return to Earth now
as if to a previous thought,
alien and out of place,
like a woman who,
waking too early each day,
finds it dark yet
and all the world asleep.
But how could my clamorous heart
lie abed, knowing all of Creation
has been up for hours?

NOTES

'Prologue'

p. 11   "birthday of Copernicus"—*The Planets* was conceived, and the prologue begun, on the 500th birthday of Copernicus.

'Mercury'

p. 21   "on the nod"—Refers to the heroin addict's nodding of the head.

'Venus'

p. 27   "shake down the planet's inhospitable air"—Carl Sagan has suggested that the Cytherean atmosphere could be made habitable. One need only inject a hardy strain of blue-green algae, which would, ultimately, increase the amount of oxygen and lower the bake-oven temperatures.

p. 27   "hymning for life-sign water and oil"—Water seems to be necessary for the growth of life.

p. 32   "her pearly tonnage . . . iron heart"—Venus always keeps the same face towards Earth each time it passes closest to our planet. I'm suggesting that, since Venus has no magnetic field and Earth has a considerable one, perhaps Venus reacts like a giant floating compass. In fact, it is more likely the gravity than the magnetism of Earth which so orients Venus.

For further bizarre elaborations of the Venus concept (specifically, in Surrealist art) see Dr. Whitney Chadwick's "Eros or Thanatos—The Surrealist Cult of Love Reexamined," *Artforum*, Vol. XIV, No. 3, Nov. 1975, p. 46.

'Earth'

p. 37   "Floppy fish . . . pond"—As primeval fish became landlocked in ponds cut off from the sea, they needed to develop a means of flopping successfully from pond to pond, lest they die as each unreplenished pond dried up. The most efficient method appears to have been feet.

'Cape Canaveral'

This poem was written after observing the launch of Viking B to Mars, in August 1975.

'Mars'

p. 72   "made gape the pine"—cf. *The Tempest*. Prospero reminds his spirit-servant Ariel that, when he found him trapped in a tree, it was his (Prospero's) "art . . . that made gape/The pine" (I.ii. 351-3).

p. 72   "Phoball"—cf. *The Cosmic Connection*, by Carl Sagan, p. 112.

p. 73   "even ice sublimes"—To transform between solid and gas without becoming a liquid.

p. 74   "thud like gekkos to the ground"—Gekkos are especially fond of clinging to walls and ceilings (a penchant facilitated by their tank-tread feet), but occasionally they eat so much that they fall down.

'Jupiter'

p. 82   "Hydrogen . . . evolved"—cf. "Saturn," p. 92. Hydrogen, the stuff of the primordial gas cloud, is the most abundant element in the universe (followed by helium, oxygen, nitrogen, carbon, and neon.)

p. 82   "chemical building-blocks lie low"—Jupiter is probably laced with organic molecules, similar to those which on Earth led to the origin of life. The coloration of these molecules may account for the hues of Jupiter's brownish belts and the so-called Red Spot.

'Saturn'

p. 91   "crêpiest ring"—A hazy dark band outlining the inner edge of the rings is called the crêpe ring.

p. 91   "A tiny moon's constant nagging"—The gravitational pull of Mimas.

p. 92   "Saturn would float"—The planet is roughly 70% as dense as water, or 5 times less dense than rock.

p. 93   "not a rack remains"—cf. *The Tempest*, IV. i. 175.

p. 95   "Paul at the scope"—The telescope is an 8" catadioptric.

p. 95   "manticore"—A legendary beast possessing a man's head, a lion's body, and a dragon's tail.

Three general notes:

1.   Astronomers tend to say "Titanian" rather than invoke the ill-fated *Titanic*.

2.   May 20th. Tonight Venus and Saturn were both 'in' Gemini, high in the sky, as if Venus were coming back to haunt me.

3.   Phoebe, like some of the Jovian satellites, rotates somewhat eccentrically—a fact commemorated in this charming poem:

> Phoebe, Phoebe, whirling high
>   In our neatly plotted sky,
> Phoebe listen to my lay
>   Won't you swirl the other way?
> Never mind what God has said,
>   We have made a law instead.
> Have you never heard of this
>   Nebular Hypothesis?
> It prescribes in terms exact
>   Just how every star should act,
> Tells each little satellite
>   Where to go and whirl at night.
> And so, my dear, you'd better change;
>   Really we can't rearrange
> All our charts from Mars to Hebe
>   Just to fit a chit like Phoebe!

(Anonymous)

'Uranus'

p. 99 "Annie Jump Cannon"—Working at Harvard, she catalogued half a million stars according to their spectra.

p. 99 "Angelina Stickney"—Wife of astronomer Asaph Hall, she prodded her husband into a few more nights' observing, during which he discovered the two satellites of Mars.

p. 100 "Caroline Herschel"—First important woman astronomer, who discovered 8 comets herself, and spent most of her life aiding her brother, astronomer William Herschel, discoverer of Uranus, in his researches and observations.

p. 100 "lying sideways . . . on bended knee"—From our perspective, Uranus is lying on its side, in the 1980's, with one pole pointing directly at the sun. This posture is unique among the planets, and probably produces unusual weather systems.

p. 101 "forty-two hearings"—One Uranian day = 42 Earth days.

p. 101 "Scheat"—Deep yellow star in the constellation Pegasus (The Winged Horse).

p. 102 "an ox is standing on my tongue"—Aeschylus, *Agamemnon*, 35-36.

p. 102 "What words . . . lips!"—This time it's Homer's *Iliad*, book 4. Flamsteed gets it wrong.

p. 104 "Edward VII"—Halley is confusing Edward VII with his notorious son.

p. 106 "all . . . are named . . . *A Midsummer Night's Dream*—Christie has good intentions, but misremembers his Shakespeare. Miranda and Ariel appear in *The Tempest*.

p. 107 "A bat flutters through the banquet hall"—In a medieval fable, life is depicted as a bird flying through a well-lit banquet hall.

p. 107 "RU Lupi"—A star in the constellation Lupus (The Wolf).

p. 107 "the man on the flying Trapezium"—Trapezium is a box of four stars in the fish-mouth of Orion's Great Nebula.

p. 107 "Mas-tab-ba-tur-tur"—A white and violet double star, known as the Little Twins, in the constellation Gemini.

p. 107 "Lick"—An observatory on Mt. Hamilton, California.

p. 108 "coal sack" —Dark region in the Milky Way at Cygnus.

p. 108　"Swan"—Constellation Cygnus.

p. 108　"Alde Baron"—Aldebaran, red giant in the constellation Taurus (The Bull).

p. 108　"Navis"—Argo Navis, a constellation originally known as the Ship of the Argonauts, and later divided into smaller parts: Puppis, Vela, and Carina.

p. 109　"at least he died . . . Böotes on"—Böotes is a constellation (The Plowman).

p. 109　"Fomalhaut"—A reddish star in the constellation Pisces (The Fish).

p. 110　"Nathanial Bliss"—Served only two years as Astronomer Royal; the reference to his only portrait being engraved on a tankard with the amusing inscription is true.

p. 112　"Paul West"—Fiction-writer and essayist, in one of whose novels, *Gala*, three characters construct a Milky Way in their basement.

p. 113　"Ophiuchus"—Constellation known as The Serpent-Holder.

p. 114　"W.H. Pickering"—A flamboyant and bizarre figure in turn-of-the-century astronomy, who was forever discovering nonexistent things, including 'platts' of insects on the moon. Of course, Herschel swore that sunspots were holes through which one could see a civilization of sun-dwellers.

p. 114　"Algol"—A white variable star in the constellation Perseus, which Ptolemy first called the "Blinking Demon." The Chinese called it *Tseih She*, "The Piled-up Corpses." And Algol becomes Goethe's "Lilis" in his *Walpurgisnacht*. A dark companion star, swinging around Algol, periodically eclipses it. The name means "ghoul," and symbolizes to astrologers the violence of the heavens.

p. 114　"Sirius . . . matter"—An intense, blue-white star in the constellation Canis Major. Sirius's dark companion is a neutron star, i.e., a star made up of what scientists call "degenerate matter" (the atoms break into shards).

p. 114　"Hyades"—Group of stars Edmund Spenser referred to as the "Moist Daughters," because the name derives from the Latin for "rain."

p. 114　"Aquila"—Constellation known as The Eagle.

p. 115   "Zubeneschamali"—A pale emerald star in the constellation Libra (The Scales).

p. 115   "Oh Be A Fine Girl Kiss Me Now"—Mnemonic for the 8 most common varieties of star spectrum.

p. 116   "a lyrical piece for electronic synthesizer"—I had in mind Mor-
□   ton Subotnick's "Silver Apples of the Moon."

'Neptune'

p. 126   "Salacia"—Mythic Neptune's wife, sometimes thought of as the Sea Queen, and sometimes as the sea itself. Here she is clearly primal Sea Queen, but later, in the footnoetic on p. 127, she becomes something closer to an embodiment of nature.

p. 129   "Shooting off . . . gone to see"—More folklore than fact. Cf. Harold C. Urey, "The Origin of the Earth," *Scientific American*, Vol. 187, No. 4, pp. 53-60.

p. 132   "wrenched from Uranus's . . . rib"—Neptune's presence in the solar system was deduced from various calculations about Uranus, then searched for and spotted.

'Pluto'

p. 139   "the fear of Antaeus"—According to mythology, Antaeus was safe only so long as he kept one hand on the Earth.

p. 139   "Eratosthenes"—A Greek mathematician and astronomer who, in the third century B.C., knew that the world was round.

p. 139   "the raucous math of Percival Lowell"—Lowell first deduced the presence of Pluto mathematically. Later, W.H. Pickering did the same. When astronomers at Flagstaff finally spotted the faint planet, they found it to be smaller than Earth and less massive. This meant that Pluto could have no sizable effect on Uranus or Neptune, and yet it was by these supposed effects that Pluto was discovered.

p. 140 "Neptune and Pluto . . . *pas de deux*"—Beginning in 1987, Pluto will move inside the orbit of Neptune and, for two decades, will no longer be the outermost planet in the solar system.

p. 141 "frontier towns on Triton"—Triton is one of Neptune's moons.

GLOSSARY

aerosol—Solid or liquid particles suspended in a gas.

albedo—Degree of reflectivity; specifically, the amount of light reflected by a body compared to the amount of light falling on it.

amino acids—Called the building-blocks of life, because from them proteins are made, and proteins are the constituents of all living things. At least, on *this* planet.

axis—An imaginary straight line about which a planet spins.

basalt—Dark, sometimes glassy, volcanic rock.

black holes—A black hole is a star which has undergone total collapse, where gravitation is so strong that even light can't escape. Black holes are much heavier than either neutron stars or white dwarfs, and so compact that, if you had a black hole that weighed as much as the Earth, say, its circumference would be only about an inch. By rotating, the black hole creates a vortex that drags things into it (like a bathtub drain). At the moment, no black holes have been indisputably identified, though many are strongly suspected. Something invisible is, obviously, difficult to spot. And though little is, or can be, known about the bizarre internal landscape of black holes (if light can't escape, neither can one's readings), the general consensus is that an object dropped in would be stretched like taffy, first in one direction, and then in the other, thus disintegrating in, perhaps, one-millionth of a second. A black hole also has a number of parts, two of which are *singularity* (the curious alter-ordinary state of physics within a black hole), and the *absolute event horizon* (the outer membrane or brink of a black hole, beyond which there is no return).

bionts—Organisms of any sort.

bute—Phenylbutazone, a narcotic.

calipers—A measuring instrument, somewhat like a compass, whose two bowed legs open out to measure round things.

carapace—Hard outer coat, as that of a beetle or turtle.

carborundum—A dark, grey-black silicon carbide.

cloud-chamber—A device for determining the paths of charged and neutral particles by examining the bubble trails they leave.

cobalt—A goblin who lives underground, damaging the silver deposits (from Middle High German *Kobalt*); also a chemical element.

crypto-biotic repose—Scientific jargon for "sleep."

Cytherean—Venerean, of the planet Venus.

DNA—deoxyribonucleic acid. A double-helix, found in cell nuclei, responsible for transmitting the genetic code. The DNA molecule is like "a spiral staircase whose handrails are sugar and phosphate" (J. Bronowski).

E. coli—A bacterium that lives in the intestines and is important for digestion.

enzyme—A protein that functions as a catalyst, making possible biochemical reactions. Enzymes are highly specific, and frequently are compared to keys (which will open only certain locks).

Euchronia—Loosely translatable as "Good Time."

glitch—A star-quake.

Gorgon—Any of three sisters (Medusa, Stheno, Euryale) in Greek mythology having snaky hair and capable of turning people into stone.

gravitation—The apparently universal mutual attraction of bodies.

flower of blood—A geyser of blood spigoted through the blow-hole of a harpooned (and dying) whale.

haggis—Scottish stew made of sheep or calf entrails mixed with onions, oatmeal and suet, and boiled in the animal's stomach.

holograph—Laser-produced, 3-dimensional images, which seem to hover in the air.

infrared—Wavelengths longer than visible light, but shorter than microwaves.

islands of Langerhans—Clumps of cells in the pancreas, which secrete insulin (a hormone controlling the blood sugar level).

Lackawanna—River in Pennsylvania.

leucorrhea—A vaginal discharge.

limonite—Yellowish-brown iron oxide.

litmus—A lichen extract that changes from blue to red when it becomes acidic, and red to blue when basic, used to indicate the degree of acidity or alkalinity.

Magellanic clouds—Two neighboring galaxies, visible in the southern hemisphere, which are satellites of our own galaxy.

magma—Pasty combo of solid and liquid; molten layer under the Earth's crust.

mare, maria (pl.)—Large dark lunar regions once thought to be seas.

meteor—The "shooting star" trail left by a meteoroid (anything from a fleck of dust to a giant tons-heavy asteroid) that, due to friction, ignites when it enters Earth's atmosphere.

mitochondrion, mitochondria *(pl.)*—Cellular inclusions responsible for the combination of oxygen with food to extract useful energy.

mojo—A voodoo doll or other object used to focus one's magic.

Nereids—Sea nymphs. Nereid is also the smaller moon of Neptune.

neutron star—An extremely dense burned-out star, about 10 miles in diameter, composed almost entirely of neutrons.

obsidian—Black volcanic glass formed when hot lava is abruptly cooled; it has an unusual tree-ring pattern, and was used by Indians to make arrowheads.

orc—Killer whale.

parsec—3.26 light years.

Pataphysics—The science of imaginary solutions.

polychotomies—Neologism on the model of *dichotomy*.

polymers—Large chains of simple, light-weight molecules.

positron—A particle with the mass of an electron but a positive charge. On collision with an electron, it destroys both, leaving only gamma rays.

prominence—A luminous cloud of gas stretching above the Sun's surface.

protein—Compounds which contain amino acids and are fundamental to an organism's growth and repair.

pulsar—A magnetized, rotating neutron star, whose radiation spurts out in regular short bursts (like a "rotary sprinkler"). They were first discovered by Jocelyn Bell Burnell, and are generally short-lived.

pumice—Light, porous, abrasive volcanic rock used in soap, among other things.

quarks—Hypothetical sub-atomic particles thought of as the ultimate units of matter (from James Joyce's *Finnegans Wake*, "three quarks for Mr. Marks").

quasar—A quasi-stellar object having a large red shift (which probably means it's exceptionally distant), moving at great speed, and emitting vast amounts of radio waves and polarized light. It's been suggested that inside of every quasar is a small cluster of giant pulsars, or a massive black hole.

radio galaxy—A galactic flux teeming with cosmic rays, and thought to include a massive pulsar-type kernel the size of a solar system, perhaps, spinning (roughly once a year) to create weak but extensive magnetic fields.

rara avis—Literally, a "rare bird" (Latin); a peculiar or unusual person or thing.

Rayleigh scattering—The preferential scattering of short rather than long wavelengths of light by molecules, giving us blue skies and reddish sunsets.

rheum—Thin, watery film over the eye.

scrimshaw—Carved whalebone or ivory.

sea-brit—Tiny crustaceans and other organisms fish commonly feed on.

shuntling—When light seems to vacillate across a shiny surface, creating effects comparable to a moiré pattern.

skink—A lizard.

surdic—A *surd* is an irrational number, like π; "utterly surdic" is the literal meaning of the word "absurd."

trilobite—Paleozoic arthropod, whose fossil remains are plentiful in upstate New York.

Urania—The muse of astronomy.

Vela-X—An X-ray source in the constellation Vela (The Sail).

weed—A horse of little value.